Skills in Advanced Biology

Volume 1

Dealing with Data

J W Garvin BSc DipEd DMS CBiol MIBiol

Head of Science, Cambridge House Grammar School for Girls, Ballymena

Stanley Thornes (Publishers) Ltd

First published in 1986 by:
Stanley Thornes (Publishers) Ltd
Old Station Drive
Leckhampton
CHELTENHAM GL53 0DN
England

Reprinted 1991

British Library Cataloguing in Publication Data

Garvin, J.W.
Skills in advanced biology.
Vol. 1: Dealing with data
student's text
1. Biometry
I. Title
519.5′024574 QH323.5

ISBN 0-85950-588-X

Typeset by Tech Set, Gateshead, Tyne & Wear
in 10/12 Century Schoolbook.
Printed and bound in Great Britain at the Bath Press, Avon.

To my children:

Clare, Neil and Stephen

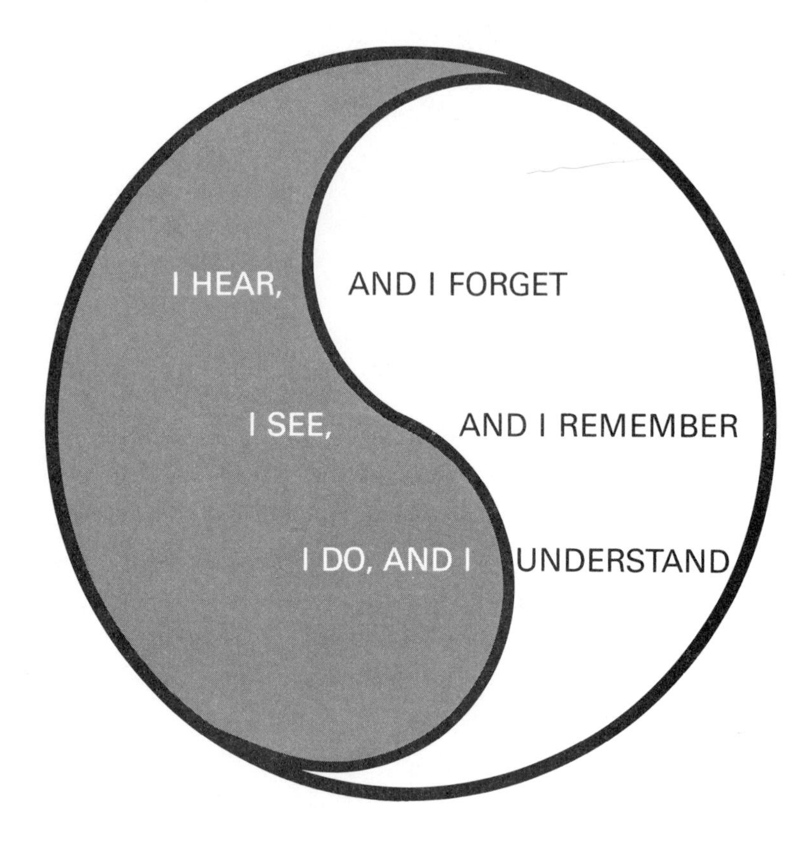
I HEAR,
AND I FORGET
I SEE,
AND I REMEMBER
I DO, AND I
UNDERSTAND

Contents

Alternative Route

The following table indicates an alternative route through this book. The order shown across the table is that given in the book. The alternative route is shown going down the table. Three topics are given, e.g.

Interdependence: 1.3.2 2.1.2 2.2.5 3.1.2 3.2.2

	Independence/ dependence	Interdependence	Frequency distributions
1.3 Relationships between variables	1.3.1 p. 4	1.3.2 p. 4	1.3.3 p. 5
2.1 Tables	2.1.1 p. 6	2.1.2 p. 8	2.1.3 p. 8
2.2 Graphs	2.2.3 p. 16	2.2.5 p. 25	2.2.6 p. 30
3.1 Graphs and parameters	3.1.1 p. 50	3.1.2 (Correlation) p. 56	3.1.3; 3.1.4 p. 61; p. 65
3.2 Probability and tests of significance	—	3.2.2 (The correlation coefficient test) p. 111	3.2.1 p. 82

ALTERNATIVE ROUTE

ROUTE THROUGH THE BOOK

Examples and Problems

Preface

To the Teacher

Many changes are taking place in G.C.E. Advanced-level syllabuses and their equivalents. Increasingly they list examination objectives which require candidates to possess a variety of skills – they need to be able to observe, record, interpret, analyse, investigate, etc. In order for students to become competent biologists and to do themselves justice in examinations they must develop these skills by careful learning and practice.

As well as these changes in the examinations, more students are staying on to study A-levels with the consequence that the ability range is widening and the variety of subjects studied with Biology is increasing.

To overcome these problems a series of skill-based active learning courses has been designed. These do not require prior knowledge as they mainly consist of procedures whereby the skills are learned using examples, and then applied using exercises. They lend themselves to individual study or as an integral part of an A level course or in combination; they are flexible so that they can fit in to any particular requirements; they can be used as an aid to teaching and as a back up and they often include testing of the concepts. The opportunity is also available for students to work at their own pace and in their own time when necessary. The range of topics has been made deliberately wide so that the examples and exercises show how the techniques can be applied in many different situations.

1 Dealing with Data

The increasingly mathematical approach to Biology, which incidentally only reflects what has been happening in the subject for a long time, can cause problems. Instead of being purely descriptive, biological observations are quantified so that relationships, trends and patterns can be determined numerically. Since biological material is notoriously variable, the application of probability theory is essential. Student groups may exhibit a wide range of numerical expertise and some students can have difficulty with the symbolism of mathematical concepts. *Dealing with Data* employs a visual approach so that students can 'see' what is meant.

The use of suitably designed Work Sheets makes answering the questions more straightforward. The Guide at the back of the Text gives all the answers so that the students can check their progress and, where necessary, work backwards through a problem thus aiding their understanding. Concerning the calculations, every step is given and explained no matter how obvious, the excessive use of symbols has been avoided but when used they have been made as meaningful as possible. All the calculations can be worked out using a calculator (with statistical functions). Practically all the examples and exercises have been taken or developed from A-level examination questions and actual studies.

Students are renowned for their ability to regurgitate facts, no matter how involved; they feel comfortable with recall-type questions. Data interpretation questions afford the examiner an opportunity to present the candidate with subject matter which is at, or beyond, the limits of the syllabus, indeed it is a better test of the candidate's analytical and interpretative abilities if the subject matter is unfamiliar. *Dealing with Data* will help students to develop those skills which are essential to answer questions involving tabular, graphical and statistical techniques. Many examination boards now include a project which requires the presentation and analysis of data.

I am a biologist and have no pretensions to being a mathematician, concerning myself primarily with using mathematics as a means by which data can be made more meaningful. *Dealing with Data* is therefore suitable for anyone wanting to know more about the application of graphical and statistical techniques in a science subject. It would be useful to teachers in other subjects, particularly those of Mathematics and Geography, and to students following introductory Biology courses at the tertiary level.

At the back of the Work Sheets there is a simple questionnaire. It would be helpful to have some feedback on how the course is working in practice.

To the Student

The day is fast disappearing when you can sit down at a book, cram in as much as possible in as short a time as possible, and hope to succeed in the examination. It is still important that you have a good background knowledge of the subject, but increasingly the trend is towards questions that test a whole range of skills – you need to be able to observe, record, interpret, analyse, investigate, etc. Initially, you probably feel very insecure with such questions because you have not developed the appropriate skills, but once you have acquired them you will feel much more confident. These skills need to be learned with understanding and then practised repeatedly. The skills will not only be essential for success in your examinations, they will also be very useful for further studies in Biology and other subjects; any employment, in fact, requires you to think, understand and interpret.

This first volume is concerned with questions that test your ability to deal with data (tabular, graphical and statistical). It has been designed to help you as much as possible with understanding the concepts. You are guided through them using a wide range of examples and exercises. You can, if necessary, work on your own, in your own time and at your own speed. If you have any difficulties, consult your teacher. The Work Sheets have been specially prepared to make the work more straightforward and they can be kept in a ring-file for reference.

At the end of the Work Sheets there is a questionnaire. If you feel inclined, please complete and post to the publishers.

All the best

JWG

Acknowledgements

The translation of ideas into book form requires assistance from many varied sources. I have been encouraged to find many that were willing to unselfishly help when necessary. I would particularly like to thank the following:

Mr R. E. Parker and Dr W. I. Montgomery of the Botany and Zoology Departments respectively, Queen's University, Belfast, for their interest and constructive advice.

Mr W. Smyth, Educational Technology Adviser, North Eastern Education and Library Board Resource Centre, for producing the trial materials.

Liam Glass for photographic assistance with the cover design.

Miss A. Graham, Principal, Cambridge House Girls' Grammar school, whose encouragement is always appreciated.

The following teachers (and their students) who took part in the trials: K. F. Adams, P. E. Anderson, J. A. Armour, G. C. Banna, R. Carlisle, T. W. Flannagan, M. Hilditch, P. Holywood, A. Jameson, M. McClean, P. M. McKernan, L. McLean, A. Magee, P. A. M. Paice, G. Taylor, D. S. Twyble.

When developing concepts and ideas it is helpful, if not essential, to have someone to act as a catalyst, or should I say enzyme! This role was admirably performed by my colleague Denmour Boyd – I would like to express my appreciation to him for all that he has done.

To all my past students who provided the stimulus for me to make what I hope is the required response.

My wife, Betty, who had to try to understand my vagaries as the ideas formed and condensed and help in preparing the typescripts; her fortitude was exceptional.

Finally I would like to express my gratitude to all those at Stanley Thornes; it was a true symbiotic relationship.

The author and publishers would like to thank the following:

Heather Angel (Biofotos) for the photograph of peppered moths (*Biston betularia*) on a lichen-covered trunk (p.51).

Phillip Coffey. Education Officer of Jersey Wildlife Preservation Trust, Les Augres Manor, Jersey, for the front cover photograph of the lowland gorilla 'Jambo'. (The cover idea was provided by the author.)

Griffin & George for information about the 'Audus' apparatus (p.25).

Arthur Guinness Son and Co (Dublin) Ltd for a photograph of W. S. Gosset (p.89) and information relating to his work from a lecture by Dr A. J. Forage delivered to a symposium on Science and Institutions in Ireland and Britain (July 8–11, 1985 at the Royal Irish Academy, Dublin) entitled 'The contribution of Guinness to the Advancement of Science and Technology in Ireland'.

The author and publishers would also like to thank the following examination boards for permission to use past questions and data:

Northern Ireland Schools Examinations Council [NI]
Scottish Examination Board (Certificate of Sixth Year Studies) [S]
Southern Universities Joint Board [SUJB]
University of London School Examinations Board [L]

1 Data

Science begins with observations. Insatiable human curiosity is the driving force behind science and we continually try to understand more and more about ourselves and the world around us. To do this we develop models or hypotheses and then try to find out if the model is a good fit with the observed facts.

Biology is particularly concerned with the way that things vary in nature; the similarities and differences between individual organisms, samples and populations. In the early days of investigation Biology was descriptive, but as time passed the descriptions by necessity became more exact and so mathematics became more and more complementary.

The information that results from observations is recorded as data – this is a strange word since it can be a singular or a plural noun. Data consist of a series of observations, measurements or facts.

1.1 TYPES OF DATA

Data are produced from two main forms of observation: natural and experimental.

1.1.1 Natural observations

These are concerned with recording similarities and differences between organisms, samples and populations. They normally involve measuring or counting.

1.1.2 Experimental observations

The second main area of biological study involves carrying out experiments to test hypotheses in order to find out, 'What happens if . . . ?'.

One of the characteristics of biological science is that the properties being recorded usually vary, so that data are produced which consist of many values – thus the properties are called ***variables***. Any single value of a variable is known as a ***variate*** and a particular group of variates a ***variate set***.

Variables can possess different properties. It is important to recognise these differences.

1.2 TYPES OF VARIABLES

Variables can conveniently be divided into three main types: qualitative, rankable, and quantitative.

1.2.1 Qualitative

These are non-numerical variables; they are descriptive, so numbers cannot be attached to them. Examples are colour, shape, behaviour, etc. Such variables are often called ***attributes*** and such observations are placed in ***categories***. These attributes or categories can be separate and distinct entities with no implication of order or preference; there is therefore no scale.

Example
A given area of chalk grassland was examined and the number of invertebrates in each of the groups shown in the table were recorded. In this example the frequencies (number of invertebrates) have been determined by counting the number of individuals within taxonomic groups. In other cases the grouping may be ecologically defined and the frequencies can be of the individuals, species or genera observed. The grouping could be geographic, e.g. countries or other clearly defined areas; or other non-numerical characteristics may be used, e.g. colour of beak. These are just a few examples of the wide range of attributes which can be used.

Molluscs	117
Woodlice	98
Millipedes	32
Centipedes	17
Beetles	30
Harvestmen	10

1.2.2 Rankable

On occasions, although the data obtained might be qualitative, there is some meaningful order present so that the individuals or populations can be ranked.

Example
When attempting to describe the distribution of a particular organism it is useful to use an abundance scale, which can be allocated values as on the left.

Rare	1
Occasional	2
Frequent	3
Common	4
Abundant	5

1.2.3 Quantitative

When an observation is described by means of a number it must have been counted or measured. The data can therefore be of two types: discrete or continuous.

Discrete

When the data are obtained by counting, the variates will only have a limited number of values, being whole numbers. They are also known as discontinuous variates.

Examples
The numbers of children in families; the numbers of petals on flowers; the number of eggs in nests; the number of heart beats; etc. There cannot be any

intermediate values or decimal places – there cannot be 1.2 children in a family or 3.7 eggs in a nest, unless of course mean values are used.

Continuous

When the data are obtained by measuring, the variates can take any value in a continuous interval (within certain limits set by the accuracy of the measuring scale). Normally the variates will possess decimal points.

Examples
The weight of students; the height of pea plants; the time it takes for plants to flower; etc.

EXERCISE
Classify each of the following data sets as qualitative, rankable, discrete or continuous.

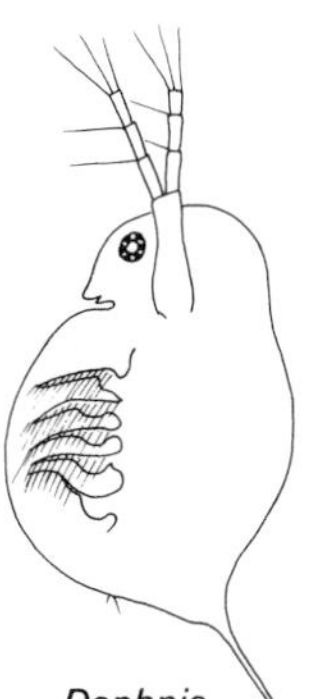
Daphnia

1 *Daphnia*, a small aquatic crustacean, was placed on a microscope slide and the respiratory movements of the abdominal appendages noted over a period of 10 s, at 17 °C. This was repeated for 10 separate animals. The results were as follows:

Respiratory movements in 10 s									
34	37	33	35	33	42	33	32	32	39

2 Sixteen broad bean seeds were weighed with the following results:

Weight of broad beans (g)							
0.91	1.45	1.62	1.15	1.84	1.43	1.28	1.30
1.71	1.63	2.11	1.47	1.12	1.61	1.74	1.70

3

	Percentages		
	On the ground	On vegetation	In the air
Araneida (spiders)	67	33	0
Hemiptera (bugs)	4	54	42
Diptera (flies)	20	37	43
Coleoptera (beetles)	7	52	41
Hymenoptera (bees, wasps)	30	26	44

4

Bean number	Length (mm)	Breadth (mm)
1	15	8
2	18	10
3	20	10
4	22	13
5	26	15
6	27	19

1.3 RELATIONSHIPS BETWEEN VARIABLES

It is important to understand clearly the relationships that can exist between different variables. In Biology a considerable amount of time and effort is expended in trying to unravel such relationships.

1.3.1 Independence/dependence

In many instances, for example when carrying out experiments, readings of two variables are obtained. If we were investigating the effect of temperature on the activity of an enzyme we would measure the temperature (in °C) and enzyme activity (amount of substrate or product). The temperature can be easily controlled and since we want to find out if there is any relationship between the two variables, we would naturally vary the temperature. Since the temperature variable is under the control of the experimenter it is called the ***independent variable*** – the values are decided by the person carrying out the experiment. The independent variable is often a physical or environmental variable (time, temperature, pH, etc). The experimenter decides at what time or at what temperature the readings should be taken.

In the above experiment enzyme activity is the ***dependent variable***, since the enzyme activity is dependent on the temperature. In such cases there is the assumption that changes in one of the variables cause changes, directly or indirectly, in the other variable. If the independent variable changes, then the dependent variable will change, but a change in the dependent variable will not produce a change in the independent variable – a change in enzyme activity will not change the temperature!

If, for example, you measured the weight of infants at various ages, which variable would be the independent one? It would be age, since you decided the ages at which to weigh the infant. If, however, you had recorded the age of the infant when it had reached certain weights, then weight would be the independent variable and age the dependent variable.

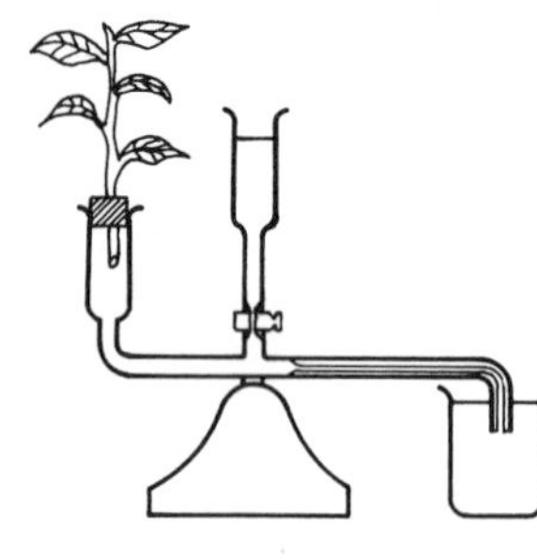

EXERCISE

Imagine that you have just carried out an experiment on the transpiration rate of a plant using a potometer. You measured the time taken for the bubble to travel 5 mm, 10 mm, 15 mm, etc.

1 Which variable is independent?

2 If, however, you had measured how far the bubble travelled in 5 min, 10 min, 15 min, etc., which variable would be independent?

1.3.2 Interdependence

Often in Biology we want to ask questions and find answers which relate to more than one variable in a sample or population. We want to find out more concerning the relationship between two or more variables – where changes in a variable are connected with changes in another variable, but there is no hint of independence or dependence, the relationship is ***interdependent***. There is no suggestion that a change in one of the variables will ***cause*** a change in the other variable. This concept of interdependence is fundamental and very widespread in Biology – that a change in one characteristic is often paralleled by a change in another; for example, longer seeds are wider, larger shells are heavier, etc.

1.3.3 Frequency distributions

Often the data collected in Biology are made up of the number or frequency of observations falling within particular categories or classes – the data are grouped.

The categories or classes may belong to any of the three types of variables that you found in section 1.2. The other variable is always the number of observations falling within the particular categories.

Qualitative

The categories or attributes are simply separate and distinct entities with no implication of order or preference.

Example
Similar to the example given in section 1.2.1, the number of invertebrates in a given area of heathland was recorded as in the table.

Molluscs	0
Woodlice	4
Millipedes	3
Centipedes	18
Beetles	68
Harvestmen	12

Rankable

Here the categories fall into an obvious and natural order.

Example
The percentage infection of sugar beet by the beet yellow virus was recorded for the years 1940 to 1947 as in the table.

1940	5
1941	2
1942	15
1943	35
1944	67
1945	65
1946	20
1947	30

Quantitative

Frequency distributions are often based on continuous variables, i.e. variables which are divided into clearly defined ranges or bands.

Example
A sample from a population of *Patina* (a small limpet) was recorded as the number of limpets falling into various size bands as in the table.

Length (mm)	Number
4.01– 6.40	30
6.41– 8.80	41
8.81–11.20	28
11.21–13.60	9
13.61–16.00	2

2 Organising and Presenting Data

2.1 TABLES – THE PRIMARY ORGANISATION OF DATA

Data should be recorded in the form of a table which has been pre-prepared. This makes the recording more straightforward and mistakes will be less likely to occur.

2.1.1 Independent/dependent variables

When carrying out an experiment all the variables except one are kept as constant as possible and this one is varied in order to determine its effect.

In an experiment to find the effect of temperature on the rate of enzyme activity, the temperature would be varied, say, from 10 to 60 °C in steps of 10 °C. For each temperature the rate of enzyme activity would be measured using a suitable technique. The temperature is thus the independent variable since it is the one that is deliberately being varied by the experimenter, and the rate of enzyme activity the dependent variable.

A table is therefore constructed before the experiment is carried out. The table will have two columns, the first containing the independent variable and the second the dependent variable.

Temperature (°C)	Rate of enzyme activity (min^{-1})

Remember that the units of each of the variables should always be given; min^{-1} just means 'per minute'.

The temperature values can be entered before the start of the experiment and as it proceeds the appropriate values of the rate of enzyme activity can be added.

Temperature (°C)	Rate of enzyme activity (min^{-1})
10	5
20	10
30	16
40	30
50	46
60	10

In most cases such a set of data is inadequate since the rate of enzyme activity should be measured at least three and preferably five times for each temperature. Such experimental repetitions are known as ***replicates*** since they are repeated as exactly the same as possible. Often class results can be pooled, each group carrying out the experiment at one or two temperatures (but a word of caution – see the test of homogeneity, p. 102).

Temperature (°C)	Rate of enzyme activity (min^{-1})					
	Replicates					Average (mean)
	1	2	3	4	5	
10						
20						
30						
40						
50						
60						

As you work through this course you will come across many examples of different experimental tables.

Tables can be quite complicated and contain a large amount of information. We can, however, come to many conclusions concerning the data from the table alone. At other times we might have to draw graphs from the data (section 2.2) or carry out a statistical analysis (sections 3.1, 3.2).

EXERCISE

The table below gives the approximate daily requirements for energy (in kJ) and protein (in g) for males and females of various ages. The typical mass for each age range is also given.

Age (years)	Mass (kg)		Energy (kJ)		Protein (g)	
	Female	Male	Female	Male	Female	Male
0–1	8	8	3850	3850	20	20
1–3	13	13	5400	5400	32	32
4–6	18	18	6690	6690	40	40
7–9	24	24	8780	8780	52	52
10–12	33	33	9200	10 040	55	60
13–15	47	45	10 460	12 550	62	75
16–18	53	61	9620	14 230	58	85
25	58	70	8780	12 130	58	70
45	58	70	7950	10 880	58	70
65	58	70	6690	9200	58	70

1 At what age is the peak energy requirement for males?

2 At what age is the peak energy requirement for females?

3 Do the figures for mass follow the trends in the figures for daily energy and protein requirement? [NI]

2.1.2 Interdependent variables

In section 1.3.2 (p. 4) you saw that on occasions we want to find out whether there is any connection between two variables, i.e. whether the relationship is interdependent. In such cases we have to make sure that the two variables measured relate to one particular organism.

Example
Shells of the sea-shore mollusc *Nucella lapillus* were measured for height and width.

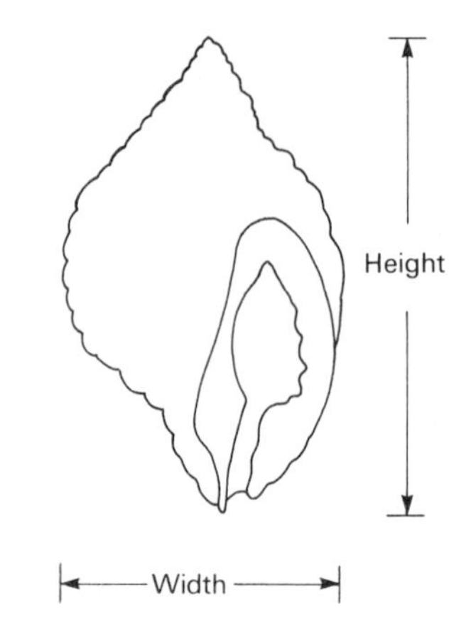

A data sheet was constructed as in the table.

Shell number	Height (mm)	Width (mm)

Each of the shells could thus be numbered and each measured in turn for the two variables.

Shell number	Height (mm)	Width (mm)
1	22	17
2	23	17
3	23	16
4	14	18
5	21	14
etc.	etc.	etc.

The two variables (or more if required) relate clearly to one particular organism and no confusion arises.

2.1.3 Frequency distributions

For frequency distributions the data have to be arranged so that the frequency of each category is known. For qualitative and discrete data there is no problem since the groups are self-definable and quite distinct from one another. When dealing with continuous variables, however, we must make sure that the classes or groups are mutually exclusive, i.e. it must be quite clear into which class each observation falls.

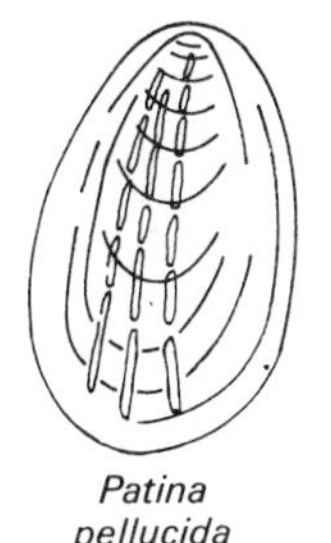

Patina pellucida

Example

The following data refer to the shell length of 110 blue-rayed limpets (*Patina pellucida*) collected from *Laminaria* (a large sea-weed) at extreme low tide level on a rocky shore. The lengths were accurately measured using an eyepiece graticule in a stereomicroscope.

Lengths of 110 blue-rayed limpets (mm)							
5.54	6.59	7.24	8.81	9.91	10.10	5.70	4.01
7.23	9.07	11.20	6.75	5.81	7.00	11.61	7.68
9.11	6.98	4.38	6.92	7.82	6.73	11.00	11.83
7.26	7.11	7.07	6.64	9.84	6.62	4.29	6.76
11.99	9.25	9.23	7.32	7.32	6.29	7.01	8.72
9.62	11.18	5.73	7.57	9.12	4.64	11.04	6.52
6.97	5.24	11.89	8.04	9.64	7.39	5.41	5.72
6.66	9.87	14.81	12.12	5.30	7.16	10.95	8.72
9.66	4.95	5.84	7.85	5.40	9.77	11.67	8.79
5.27	6.60	7.67	8.45	10.33	4.45	5.82	6.03
8.45	9.57	12.04	8.12	5.02	6.65	5.36	8.49
10.12	8.08	4.75	6.55	9.70	12.39	6.21	7.98
5.81	10.42	4.39	6.15	10.25	13.20	6.10	6.65
9.16	10.71	5.43	6.31	9.53	13.80		

This mass of data is relatively meaningless as it stands. What is required is to group the data into various classes defined by certain length bands.

First, the number of classes has to be determined. This can be difficult, since it can depend on the nature of the data, but a good 'rule of thumb' is that the number of classes should be 5 times the log of the number of observations. In this case there are approximately 100 observations, so the number of classes should be $5 \times \log 100 = 5 \times 2 = 10$.

Second, the class size has to be determined. To do this the smallest value is subtracted from the largest value and the result is divided by the number of classes required minus one. In the example the longest limpet was 14.81 mm and the shortest 4.01 mm, so $14.81 - 4.01 = 10.80$. This value is then divided by $10 - 1 = 9$, so the class size is $10.80/9 = 1.20$.

The classes now have to be clearly delineated so that there is no possibility of confusion as to whether a particular limpet falls into any given class.

Starting at the shortest limpet, $4.01 + 1.20$ (the class interval) $= 5.21$, so the first class is 4.01 to 5.21 mm. The next class will be 5.21 to 6.41 mm. If a limpet was 5.21 mm long, however, should it be placed in the first or second class? The first class can be defined more clearly as 4.01 mm up to, but not including, 5.21, the second as 5.21 mm up to, but not including, 6.41, etc. We can either use the convention (probably the most common) class 1 = 4.01–5.20; class 2 = 5.21–6.40 etc., or we can use the sign meaning up to but not including, so that the classes would read $4.01 < 5.21$, $5.21 < 6.41$, etc.

The classes would therefore be as follows.

4.01– 5.20	or	4.01 < 5.21
5.21– 6.40		5.21 < 6.41
6.41– 7.60		6.41 < 7.61
7.61– 8.80		7.61 < 8.81
8.81–10.00		8.81 < 10.01
10.01–11.20		10.01 < 11.21
11.21–12.40		11.21 < 12.41
12.41–13.60		12.41 < 13.61
13.61–14.80		13.61 < 14.81
14.81–16.00		14.81 < 16.01

This gives the required 10 classes. All that has to be done now is to find out how many limpets fall into each of these classes. It is always better in such cases to have some system so that no confusion arises. The best method is to use a ***tally chart*** such as the one given on Work Sheet 1.

EXERCISE

Complete the tally chart (Work Sheet 1) in the following manner. The first limpet measured 5.64 mm so it falls into the 5.21–6.40 mm class. A line is therefore put in the second box down and close to the left-hand side of the '5' column, and a tick is placed beside limpet 1 in the table. It is important to tick off each observation as you record it otherwise you will easily get mixed up. The second limpet (going down the column) measured 7.23 mm, so it goes into the 6.41–7.60 mm class. A line is drawn in the appropriate box and limpet 2 ticked off. This process is continued; it has in fact been completed down to limpet 17 (length 6.98 mm). This limpet was the sixth one in class 6.41–7.60 mm. Note that the fifth tally mark is drawn diagonally through the previous four lines. The sequence is as follows:

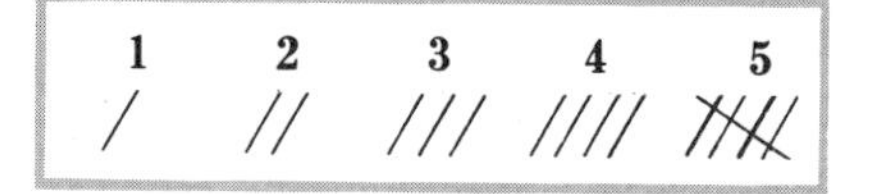

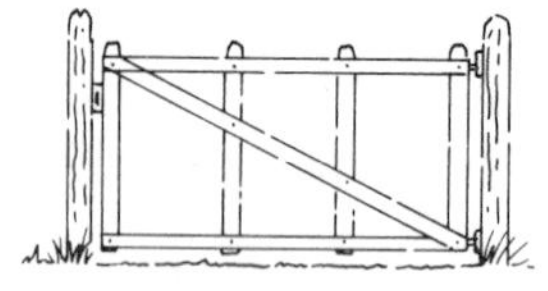

This is often called 'gate' scoring, presumably since the five lines look like a gate.

Continue this tallying process until all the limpets have been assigned to a class. Using the tally chart it is then a simple matter to complete the frequency column, giving the total number of limpets in each class.

1 What are the frequency values for each of the classes?

The reason for grouping the data is that a pattern is formed by the changes of the frequency value as the values of the classes change. You will have seen that a pattern already emerges on the tally chart.

2.1.4 Contingency tables ($n \times n$ tables)

In particular cases we might want to examine organisms for two characteristics and to classify each characteristic into a number of categories. The data can be set out as an ***array*** or ***matrix***, thus making analysis more straightforward.

2 × 2 tables

The 2 × 2 table is the simplest type of contingency table and occurs when two sets of observations have two categories each so that there are four possible combinations. The categories of one set of observations are labelled along the top and those of the other on the left-hand side. The frequencies are then entered in the appropriate boxes or cells. The totals of the columns are added along the bottom and the totals of the rows on the right-hand side. At the bottom right-hand corner the grand total is entered:

		First category		Row totals
		A	B	
Second category	A	cell 1	cell 2	
	B	cell 3	cell 4	
Column totals				Grand total

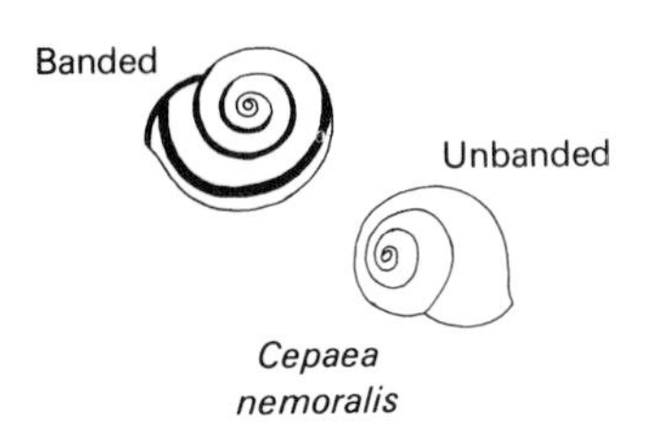

Example
The banded snail, *Cepaea nemoralis*, was collected from two localities – a beechwood and under hedges. Of 150 snails collected from the beechwood 40 were found to be banded (dark bands on the shells) and the rest unbanded. Of 70 snails from under hedges, 45 were banded and the rest unbanded.

The data are entered in the table as follows:

		Type of snail		Totals
		Banded	Unbanded	
Location	Beechwood	40	110	150
	Under hedges	45	25	70
	Totals	85	135	220

c × *r* tables

There can be any number of columns and rows in a contingency table. You will be looking at examples of such arrangements later.

2.2 GRAPHS – THE SECONDARY ORGANISATION OF DATA

Graphs are very much a part of everyday life. They appear regularly in newspapers, magazines and on television. They can record a considerable amount of information in a small space and give the observer a picture of facts and relationships which could not be understood so clearly, if at all, from verbal or tabular forms. Almost any numerical data and many that are not can be represented by a graph.

As you have already found out, a major problem for biologists is the inherent variability of the material, which is compounded by random error introduced by the observer or experimenter. Graphs can help us to visualise this variability, but still enable us to perceive trends. In many of the problems given in this book this variability has been reduced so as not to confuse matters. There are many different kinds of graphs and this book attempts to cover many of them, so that you will become aware of the usefulness of the graph in visualising the data that are presented.

Skill in the use of graphs requires practice, like the development of any skill. It is no use just reading about graphs, you have to draw them. As well as being an excellent means of presenting data, graphs are very useful when it comes to analysing data, as you will find out in section 3.

2.2.1 Drawing graphs

The ability to draw graphs properly is a skill that is essential to all Biology courses. The development of this skill is often neglected, but like other skills it can be acquired through knowledge and above all by practice. In this section we will be concerned only with line graphs, but most of the principles explained here apply also to other types of graphs.

If graphs are to be used, then they should always be used to the best advantage. You should always aim to construct the best graph that you can. Not only should the following points be taken into account, but every graph you draw should be constructed as carefully and neatly as possible. Always draw graphs in pencil and complete them in ink or ball-point pen.

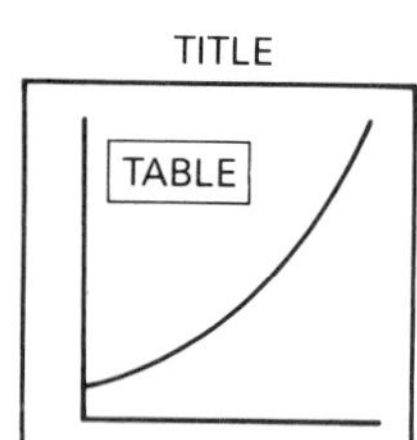

The title

Every graph should have a title along the top to specify clearly what it represents. You may draw a graph of the results from an experiment, but if it has no title then it can be relatively meaningless and quite useless for future reference. Where possible, the data should also be attached to the graph; better still, the table can be added to the graph (if it does not interfere with the plotting).

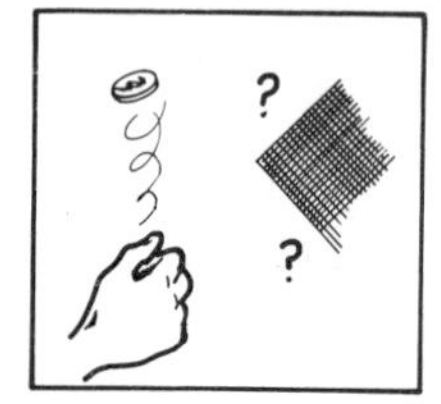

The axes

Many students often choose the wrong axes when constructing graphs. Even more of them express the opinion that the choice of axes is arrived at purely by guesswork, chance or even intuition!

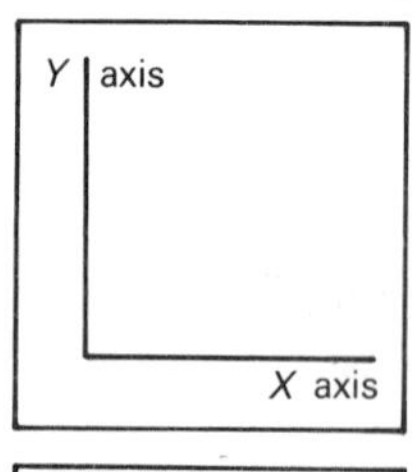

If there are two axes, they are commonly known as the horizontal and vertical axes. Since the so-called 'vertical' axis doesn't rise vertically out of the page, it is better (and more mathematical) to call them the X and Y axes. The X axis goes across the page (remember X is a-cross) and the Y axis goes up the page. The X axis is known as the ***abscissa*** and the Y the ***ordinate***. The X axis is reserved for the ***independent*** variable and the Y axis for the ***dependent*** variable. Both are called variables since they can have different values on different occasions. The two axes meet at the ***origin*** which is denoted by O.

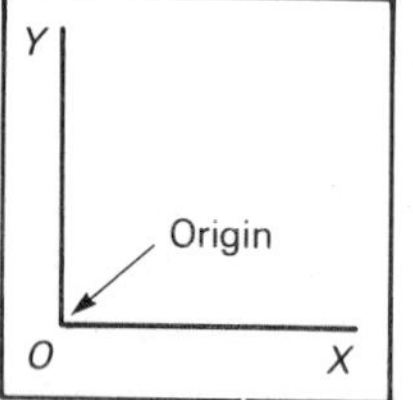

In section 2.2.5, p. 25 you will be looking at the situation when the two variables are interdependent.

There is a slight complication when time is one of the variables. Time is traditionally plotted on the X axis whether or not it is the independent variable – this is known as a time series graph. It is better, however, to follow the rule given above that the independent variable always goes on the X axis, so that no confusion arises.

Positive values of X move to the right of the origin, negative values to the left. Positive values of Y are above the origin, negative values below. The quantities of the variables increase as they progress away from the origin.

Generally it is better to have the origin at point 0,0 i.e. $X = 0$ and $Y = 0$, but consider the following situation:

X	**60.0**	**65.0**	**70.0**	**75.0**	**80.0**
Y	**20.0**	**22.5**	**25.0**	**30.0**	**35.0**

When these variables are plotted with the origin at $X0$, $Y0$ the result is as follows.

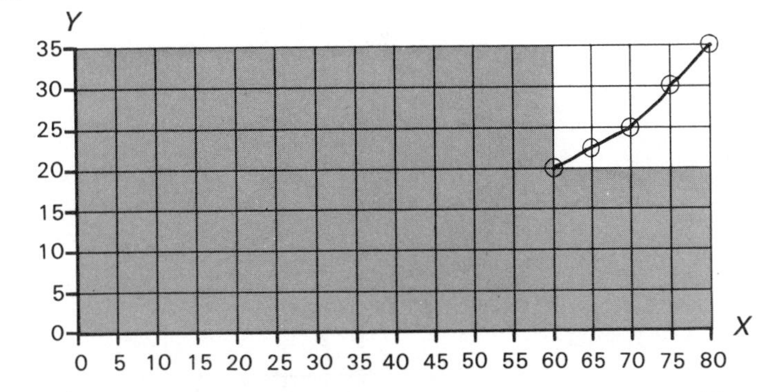

You can see that the points are bunched together over a very small range some distance from the origin. The complete shaded portion of the grid is unused and accuracy suffers. In such cases the points should be spread over the whole of the grid by a suitable choice of scales. To do this ***displaced origins*** are used. Always indicate that the origins are displaced by using broken lines as shown in the graph below.

Note, however, that there are special occasions when proper origins have to be used. You will be looking at these later.

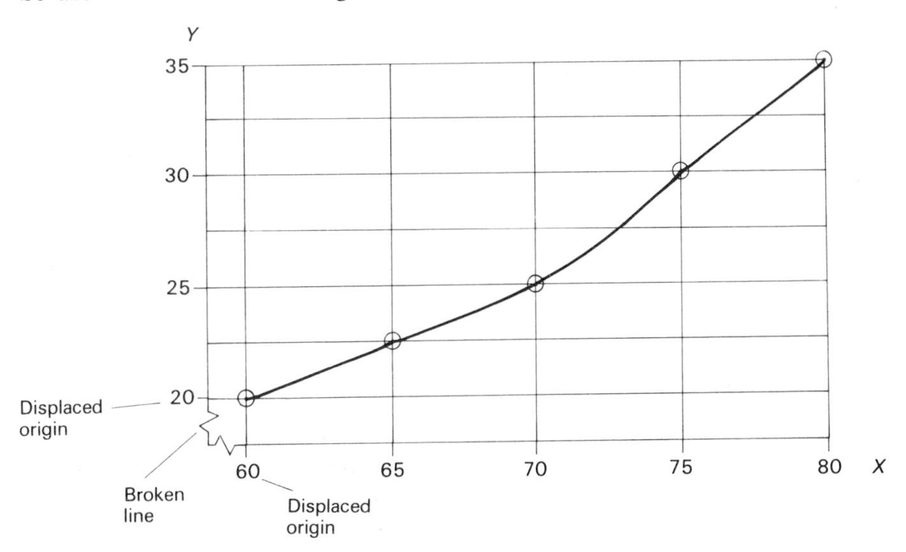

Suitable scales must be determined for both the X and Y axes. First you should place the origin at a point which will leave space both on the left and along the bottom for the scales and labels. Unfortunately most published grids leave only a very thin margin around the edge, which is quite inadequate. Always choose simple scale divisions in order to reduce the effort involved in estimating fractions of a division. Each small division should be

1, 2 or 5 times a power of 10, e.g. 0.1, 0.2, 0.5, 1, 2, 5, 10, 20, 50, etc. Never choose a scale based on one small square being 3 or 4 times a power of 10 since it will be difficult to plot the points accurately. Even if it means reducing the size of the graph, always follow this advice.

WORK SHEET 2

EXERCISE

1 The *X* axis below has to be divided into a scale which will read from 0 to 10. What would be the best way of doing this, taking all the factors given above into account? (see Work Sheet 2, question 1).

Sometimes it is easier to fit both the scales if you turn the grid around. If, for example, the *X* scale went from 0 to 8, and the *Y* scale from 0 to 6, turning the grid sideways might enable the scales to fit better.

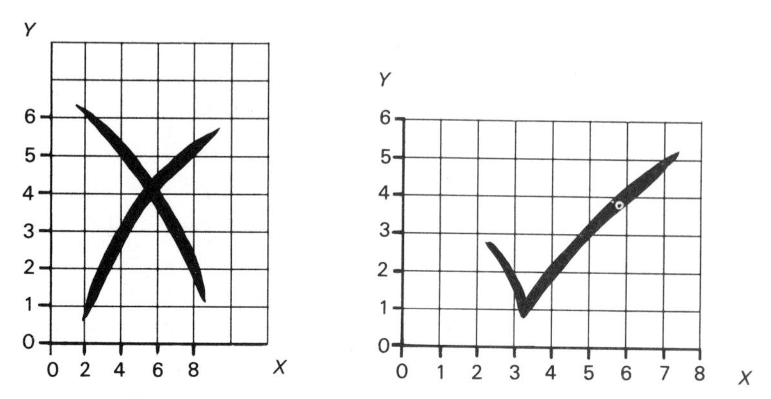

When you are plotting data from an experiment, you should make sure that the accuracy of the data is reflected by the accuracy of the scale reading. If the scale is too small then the care in obtaining the data would be wasted, and if the scale is too large, a false appearance of accuracy could be given.

The points

Any point on the grid can be located by giving it an *X* value and a *Y* value; these two values are known as the ***coordinates*** of the point. The original method of locating points was with a fine pin. If we substitute a ***very fine*** pencil point for the pin, it can be just as difficult to see, and so it is surrounded by a small neat circle. This is visually preferable to the cross. Remember not to use only dots for line graphs since they can be confused with odd marks on the paper and they get obliterated when the graph line is passed through them.

Interpolation

Joining up the points that have been plotted is termed ***interpolation***. If the points lie on a straight line use a ruler. Often the points only approximate to a straight line; variations may be due to the variability of the material or to experimental error, i.e. the sum of equipment error and observer error. In such cases, position a transparent ruler so that as many points as possible lie on the edge of the ruler; those which don't should be roughly equally distributed on both sides of the line.

2 Using a clear plastic ruler draw straight lines through the points on Work Sheet 2, question 2.

When we have a set of points that appear to follow a continuously curved line, then interpolation is much more difficult.

3 Using a fine pencil draw in a curved line on Work Sheet 2, question 3, making sure that it passes through all the points.

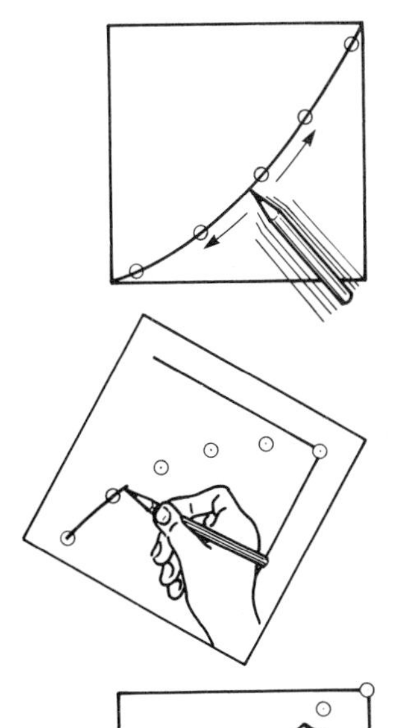

Most people find this quite difficult to do accurately and neatly and the result is usually messy.

Possibly you drew the line with the pencil held in the position shown in the diagram, and the line consisted of a series of short lines that only approximate to a true interpolation. It would be impossible to obtain intermediate readings from such a curve.

A much better method is to turn the page around so that the points lie on an arc drawn naturally by rotation of the wrist, i.e. the hand is on the concave side of the curve.

Do not draw the line using short interrupted strokes since this does not give a neat continuous line. Practise drawing the line without letting the pencil touch the paper. Rotate your wrist using the 'heel' of your hand as a fulcrum. When you think that you have it right, draw the line in one continuous movement. (The position of the paper will be virtually the same for left-handers). This is quite difficult at first, but you should improve with practice. You may find it more convenient to hold the pencil steady with one hand and rotate the paper with the other hand, again using the heel of the hand as the point of rotation. With difficult lines it may even be necessary to rotate both the wrist and the paper at the same time.

The points in the diagram on the left represent a curve which you will come across occasionally in Biology. This type of line should be drawn in two stages. The first stage (numbered 1 in the diagram) is drawn with the paper turned upside down, and the second stage (numbered 2) with the paper the right way up.

The type of curve shown here is also drawn in two stages; the paper for the second stage being rotated 90° relative to the position for drawing the first stage.

EXERCISE

1 Join up the various sets of points given on Work Sheet 3 using the suggestions given above. Keep trying until you get them right.

Extrapolation

WORK SHEET 3

By extrapolation we mean the continuation of the interpolated line beyond the range of the plotted points. This enables us to predict what might happen outside the range of measured values. The extrapolation should always be drawn as a dotted line so that it contrasts with the continuous line of the rest of the graph.

At times extrapolation may be justified, but it can often be totally misleading.

WORK SHEET 4

EXERCISE

The graph (log/linear) given on Work Sheet 4 was drawn from values obtained during an experiment to investigate the action of an enzyme at different temperatures.

1 Interpolate and then extrapolate the points to estimate the yield at 50 °C.

2.2.2 Linear grids

The linear grid is more commonly used than any other type – in fact it may well be the only type of graph paper that you have come across.

The grid is made up of equal divisions on both the scales, i.e. they are both arithmetic. The smallest unit of the grid is usually 1 mm, the 5 mm and

10 mm lines being more heavily ruled for convenience. Grids without any heavily ruled lines can be obtained – these are known as quadrilles. Linear grids can also be obtained with scales divided into other units, e.g. 2, 10 and 20 mm.

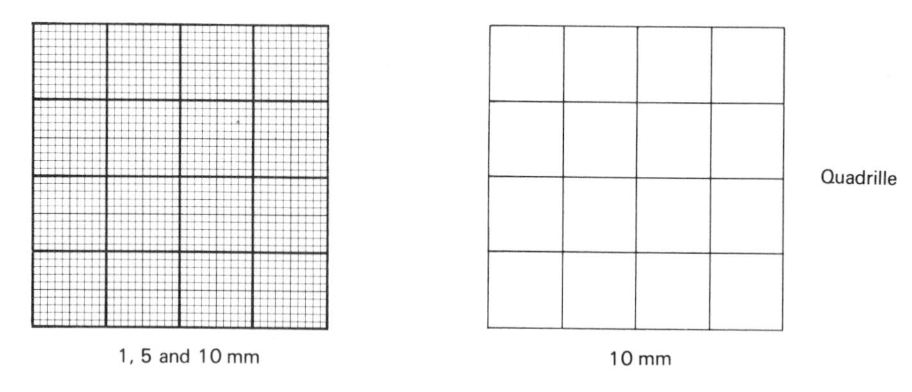

2.2.3 Line graphs

Line graphs are the most common and in many ways the most important type of graph that you will come across. Not only are they useful in displaying information, but they can also be used to solve problems. They can be drawn quickly and accurately, and allow a large number of values to be plotted in a compact space. A number of graphs can also be drawn on one grid to allow visual comparisons to be made.

Single line graphs

In these cases only one line is drawn on the grid. This shows clearly the relationship between the X and Y variables.

Jagged line

Unless accurate interpolations are required it is not necessary to go to the trouble of joining up the points by smooth curved lines. Adjacent points can be connected by short straight lines, thus giving the graph a jagged appearance. Sharp peaks often appear on the graph when in reality they may not actually occur.

The X variable is often time, the readings being taken at discrete intervals (minutes, hours, days, months or even years). A typical example of a jagged line graph is the temperature chart found at the bottom of a hospital bed. In the example below the patient's temperature was taken at four-hourly intervals.

Temperature chart

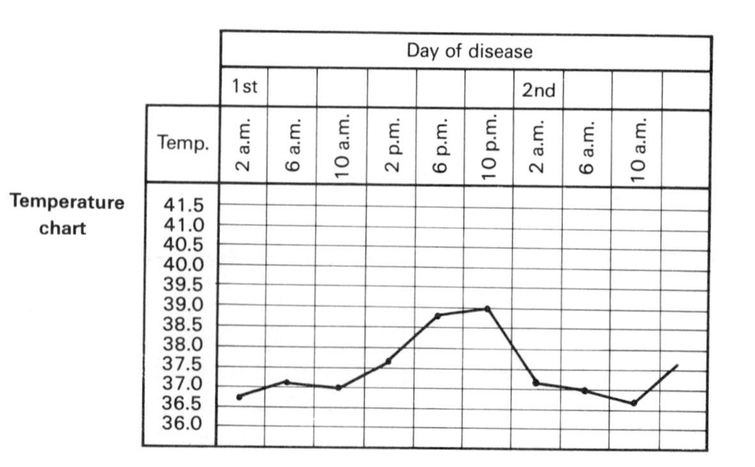

WORK SHEET 5

EXERCISE

The blood sugar level of a normal subject was measured hourly during an 18-hour period (6 a.m. to midnight). Meals were taken at 7 a.m. (breakfast), 1 p.m. (lunch) and at 4 p.m. (tea). The results are given on Work Sheet 5.

1 Draw a graph of the data on the grid provided.

2 From the graph what do you think is the approximate norm for the blood sugar level?

3 What is the effect on the blood sugar level of eating a meal?

4 Why does the blood sugar level drop between meals?

Curved line

When the plotted points appear to lie along a smooth curve (allowing for experimental error) then interpolation is justified and intermediate points may be read off the graph. In such cases joining up the points by means of a curved line is necessary and correct.

EXERCISE

The growth of a single cucumber leaf was determined over a 20-day period by measuring the area of the leaf. The results were as shown in the table.

Days	1	2	3	5	7	9	10	11	13	15	17	18	19	20
Area of leaf (cm^2)	3	5	10	30	55	90	120	150	200	220	230	234	236	236

The 'number of days' scale occupies the X axis since this is the independent variable (the experimenter decided on what days the leaf would be measured) and the leaf area scale occupies the Y axis.

The points have been plotted for you on the grid on Work Sheet 6.

1 Interpolate the points carefully with a smooth curve, remembering the instructions given in section 2.2.1.

Let us now find out how intermediate values are determined from the graph.

2 What was the area of the leaf on day 6? On the graph that you have drawn, project a line up from the X axis at day 6. Where it intersects the graph line project a line across to the Y axis to give the value of the leaf area required. Actually draw these lines on the graph; it is a good habit to get into!

3 What was the area of the leaf on day 12?

4 On what day was the area of the leaf 70 cm^2?

5 On what day was the area of the leaf 210 cm^2?

6 Between which days was the rate of growth fast?

7 Between which days was the rate of growth slow?

Compound line graphs

Several graphs may be drawn on the same grid. The greater the number of lines on a grid, however, the more difficult it becomes to identify the important features of any particular variable. If the various lines don't occur on the same part of the grid and don't overlap, then many lines can be included without causing any confusion.

Occasionally we may want to compare a number of different variables. Rather than drawing each line on a different grid it is better to put them all on one grid, thus allowing easier visual comparisons. In such cases the points and lines have to be made visually distinct from one another. Different types of line can be used, e.g. continuous line, dotted line, dot-dash line, etc. This effect can be enhanced if the points are located using different symbols, e.g. ○, ×, □.

In certain circumstances two or more Y (and even X) scales may be required on the same grid. Such a situation arises when we need to graph variables which are expressed in different units. Never put more than one Y scale for the ***same*** units since this can be very misleading. The scales can easily be arranged so that the lines of the graph are close together or far apart.

Two lines: one *Y* scale

EXERCISE

WORK SHEET 7

Twelve randomly selected soil samples from a 30 m^2 area of grassland were thoroughly examined and the mean number of mites and collembolans (spring-tails) in each sample was recorded. The same procedure was repeated at various times during the year.The results are given on Work Sheet 7.

1 Plot the results on the grid provided in order to illustrate them as clearly as possible.

2 Suggest reasons for the variation in mean numbers for both mites and collembolans.

A collembolan

WORK SHEET 8

The table on Work Sheet 8 records the percentage saturation of the blood with oxygen in a foetus and its mother.

1 Plot these results on the grid provided.

2 In what way are the two curves similar?

3 In what way are they different? Explain why this is so.

WORK SHEET 9

Work Sheet 9 gives results from experiments that were designed to determine the effect of the concentration of auxin on the growth of shoots and roots of oat seedlings. A series of concentrations were applied to their surfaces and the percentage stimulation of growth relative to a control, whether negative or positive, recorded.

1 Plot the data on the grid provided in order to show the results as clearly as possible.

2 What conclusions can you come to regarding the responses of seedling shoots and roots to the auxin?

Two lines; two *Y* scales

Sometimes two Y axes are required because there are two dependent variables.

EXERCISE

In the following problem, two quite different organisms, a goose and bamboo, were measured for growth over a period of time. The growth of the goose was determined by an increase in weight whereas in the bamboo the height was measured. The results are given on Work Sheet 10.

1 Draw two graphs of the data on the grid provided, using two Y axes.

2 Discuss the significance of any similarities or differences shown by the graphs.

Three or more lines; any number of *Y* axes

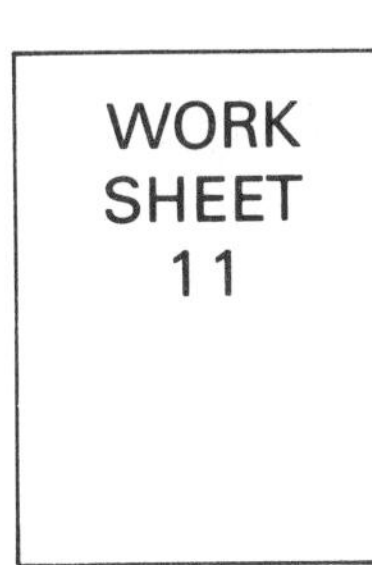

EXERCISE

The table on Work Sheet 11 is a summary of the effects of dredging for the common cockle (*Ceratoderma* [*Cardium*] *edule*), a bivalve mollusc living in the bed of a river estuary. The figures show the influence of continuous and of short-term dredging on the percentage survival of the cockles during their second growth season.

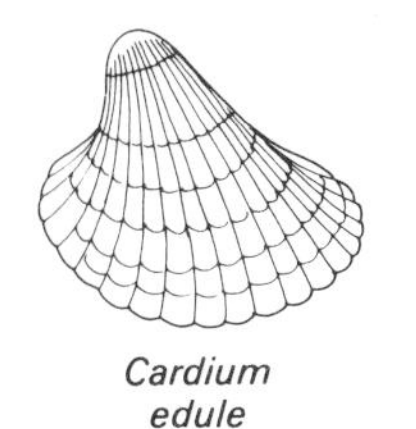

Cardium edule

1 Plot the data on the grid provided, to compare the relative survival of the cockles under the different circumstances indicated above. (Since there are three lines, instead of using different types of lines, you could try using different colours. Make sure that the coloured pencils are well sharpened.)

2 What conclusions can you draw about the effects of these fishing methods on the percentage survival of the cockles?

3 What advice about dredging methods could you give to the authorities concerned with maintaining stocks in the cockle beds? [L]

When a number of lines have to be compared graphically and there are a number of dependent variables to be considered, either the lines can be superimposed on top of one another, or the graphs can be placed above each other. In the following example we will try both of these methods to find out which is the most successful.

WORK SHEET 12

A sluggish overweight person was given a 5-day course of thyroxin; the levels of three metabolic functions – pulse rate, body mass and basal metabolic rate (BMR) – were measured during the treatment and for the following three weeks. The results are shown in the table.

Days	Pulse rate (beats min^{-1})	Body mass (kg)	BMR (per cent of normal)
0	70	66	55
2	70	66	55
4	70	62	80
6	90	63	85
10	80	61	105
14	80	61	100
18	75	60	90
22	70	59	85
26	68	58	82

[NI]

1 Plot these data on each of the grids (A and B) provided in Work Sheet 12.

2 Which seems to be the best?

Three-dimensional line graphs (optional)

WORK SHEET 13

Although three-dimensional graphs are much more difficult to construct than ordinary graphs on linear paper, they are useful for particular purposes, and are visually very appealing.

An experiment was set up to find the effect of varying both the light intensity and the carbon dioxide concentration on the rate of photosynthesis. The results are given on Work Sheet 13. From what you have already learnt in the previous section you could draw a compound line graph as on the next page.

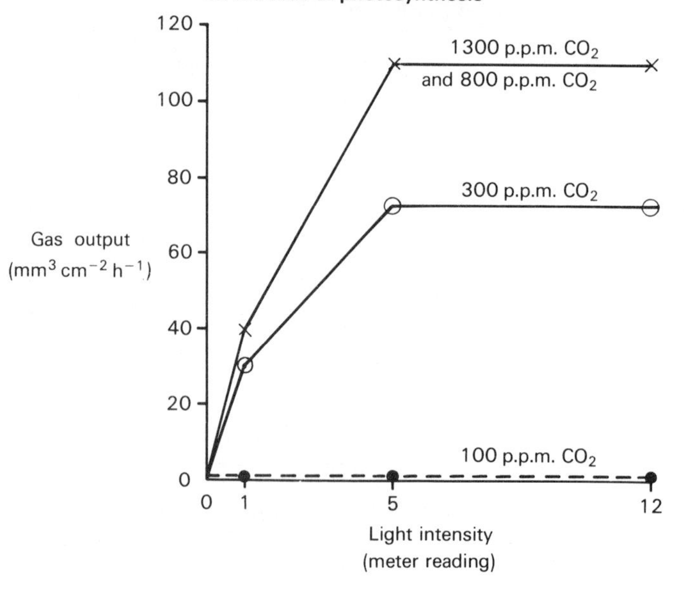

Although we could extract the information from this compound line graph, it does not show clearly the relationships between the three variables.

There are, in this case, ***two independent*** variables; the experimenter carried out photosynthesis at particular light intensities ***and*** in different CO_2 concentrations. Where there are two such independent variables and one dependent variable, the relationships can best be illustrated by means of a three-dimensional graph.

A three-dimensional graph of the data would look like this:

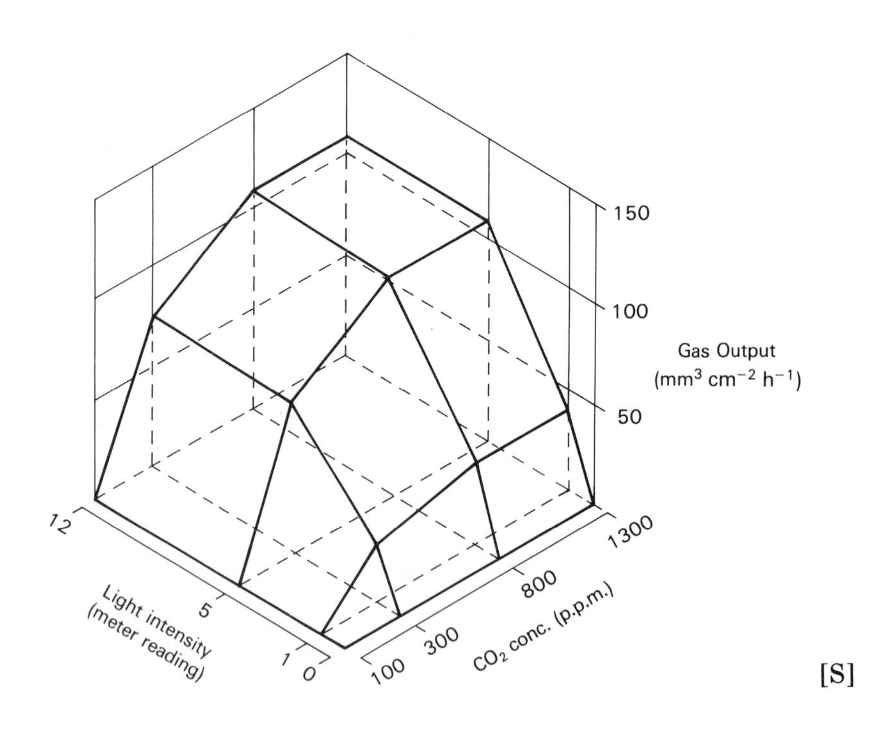

[S]

In order to find out how three-dimensional graphs like this one are constructed, follow the instructions given below. Use the grid provided on Work Sheet 13.

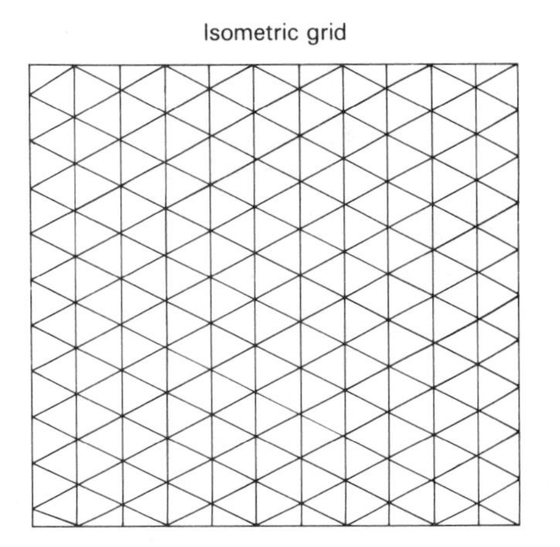

The axes

Since there are three variables there are three axes, which we will call the X, Y and Z axes. The X and Y axes are allocated to the independent variables and the Z axis to the dependent variable. In this example we will put 'light intensity' on the X axis, 'CO_2 concentration' on the Y axis, and 'gas output' on the Z axis.

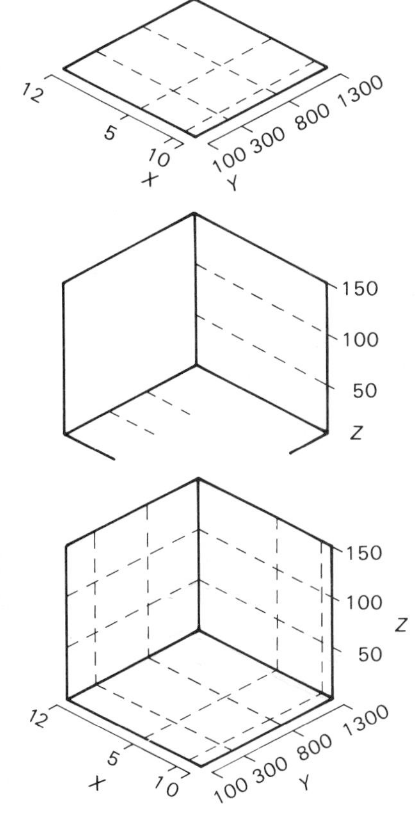

Consider the base of the three-dimensional graph as an ordinary linear grid. When using the isometric grid (see above) it looks as if it is sitting 'flat' on the page. Draw this base on the isometric grid (Work Sheet 13) using a pencil. It has been started for you. In order to simplify the plotting later on, project dotted lines from the units on the X and Y axes as indicated in the diagram on the left.

Now complete the Z axis, and project dotted lines as shown on the left at Z values of 50, 100 and 150.

Continue the dotted lines on to the other faces.

Your completed grid should now be as shown. (You should notice that this grid has the habit of reversing i.e. it sometimes 'sits out' from the page like a solid cube rather than 'going into' the page.)

Plotting

Each point is designated by the three values of the three variables, taking the order of the variables as X, Y and Z, e.g. the first box in the table is light intensity 0, CO_2 concentration 100, and gas output 0, so the point is 0,100,0. Where is this point on the grid? Examine the following diagrams.

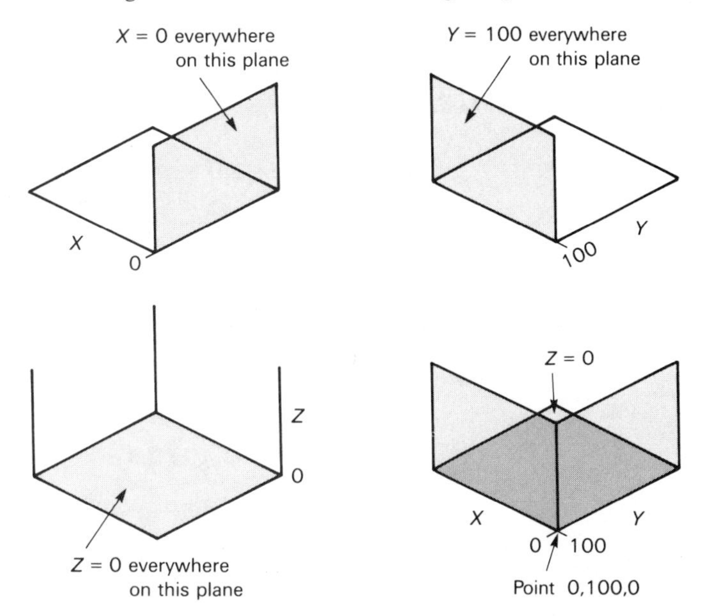

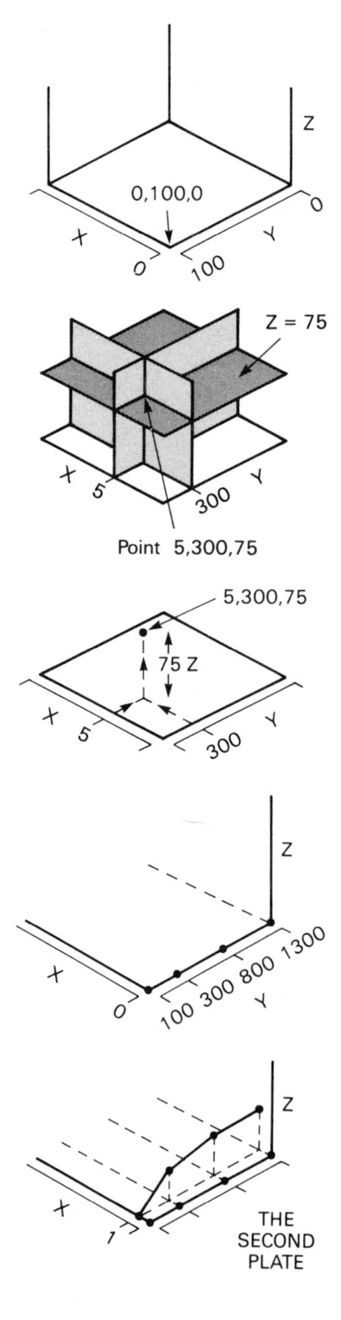

Locate point 0,100,0 on the grid.

For light intensity 5 and CO_2 concentration 300, the gas output is 75. Try and find out where point 5,300,75 would be on the grid. The following diagrams should help you.

To locate this point you have first of all to find the $X = 5$ position; move along this line until you come to the intersection with the $Y = 300$ line; finally, count up from this point 75 units along the Z axis.

To prevent confusion between the located points it is better to construct the graph in plates. Consider the light intensity = 0 values first, i.e. the first column in the table. The point values are 0,100,0; 0,300,0; 0,800,0; and 0,1300,0.

Locate these points on the grid, dot each one, and then join them up with a straight line (see diagram).

Now go to the second column, light intensity = 1. The point values are 1,100,0; 1,300,30; 1,800,40; and 1,1300,40. Locate these points, dot them and join them up with short straight lines (see diagram). This completes the second plate.

The third column contains the values for light intensity 5, and the fourth for light intensity 12.

EXERCISE

1 What are the point values?

Carry out the same procedure as above for the third and fourth columns.

Now join the plates together by short straight lines for the different Y values.

Finally complete the graph in ink so that it looks similar to the one shown on page 20.

2 Describe the effects of light intensity and carbon dioxide concentration on the rate of photosynthesis.

3 What is the volume of gas produced when the light intensity is 4 units and the the CO_2 concentration is 500 p.p.m?.

4 What is the volume of gas produced when the light intensity is 4 units and the CO_2 concentration is 1300 p.p.m.?

2.2.4 Logarithmic graphs

In many biological situations the ***rate*** of increase or decrease of a variable may need to be established, e.g. the rate of increase of a population or the rate of breakdown of a substrate by an enzyme. We saw in the last section that linear graphs show changes in quantity, where actual ***amounts*** of increase or decrease are compared. If such graphs are used to plot rates few conclusions can be drawn, and they can be misleading. For particular tasks there is no substitute for the logarithmic graph. Their advantages should be understood and they should be used more often in biological investigations.

Logarithmic grids

In this type of grid one or both of the scales are divided according to the logarithmic progression.

The logarithmic scale consists of a range of ***nine*** intervals spaced out by distances which are in proportion to the logarithms of 2, 3, 4 . . . 9 (base 10). The positions, when drawn, appear as below.

1 2 3 4 5 6 7 8 9 10

As you can see there is a regular 'bunching' of the lines every 10 points. Such a 'decade' of values is known as a ***cycle***, and each cycle has interesting and important properties: the distances between points 1 and 2, between points 2 and 4, and between points 4 and 8 are all the same. Doubling will therefore produce the same distance.

The logarithmic grid is defined by the number of cycles it possesses, i.e. one cycle, two cycles, three cycles, etc. The grid can have as many cycles as required, but normally only up to three cycles are used.

Log × linear grid

In this type of grid one axis is ruled arithmetically and the other logarithmically. It is also known as semi-log, single-log, one-way, log/normal or log-ratio paper.

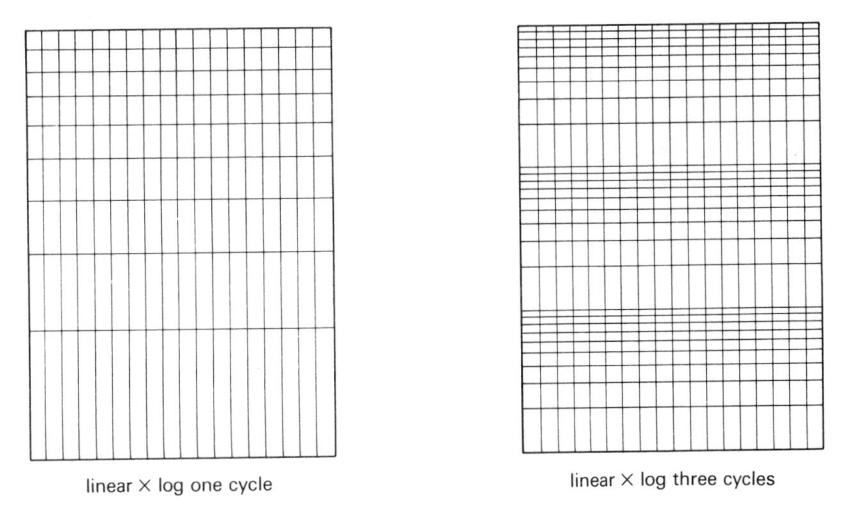

linear × log one cycle

linear × log three cycles

Log × log grid

In this case both of the axes are ruled logarithmically and each axis can have any required number of cycles. It is also known as log/log, double-log or two-way paper.

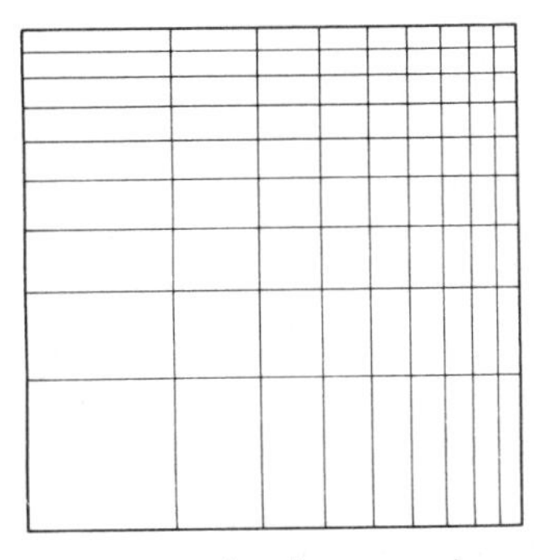

log one cycle × log one cycle

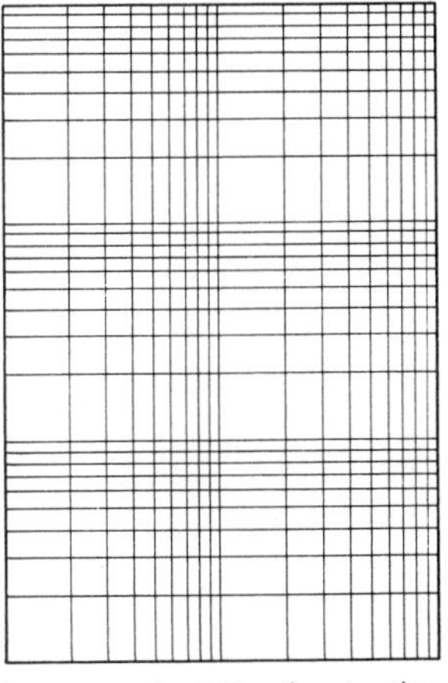

log two cycles × log three cycles

The number of cycles

Under what circumstances do we use a particular number of cycles? This depends on the ***range*** of values for the variable(s) under consideration.

The end points of a cycle do not need to be '1' and '10'. As long as the upper value is 10 times the lower value then any range can be used, i.e. the end points of a cycle could be 10 and 100, 100 and 1000, 1000 and 10 000, etc. If the range is 1 to 10 then the intermediate points will be 2, 3 . . . 9; if 10 to 100 they will be 20, 30 . . . 90, etc. The end point can also be less than one, e.g. the range could be from 0.1 to 1, when the intermediate values would be 0.2, 0.3 . . . 0.9, or 0.01 to 0.1 and the intermediate values 0.02, 0.03 . . . 0.09, etc. The following diagrams should make this clearer.

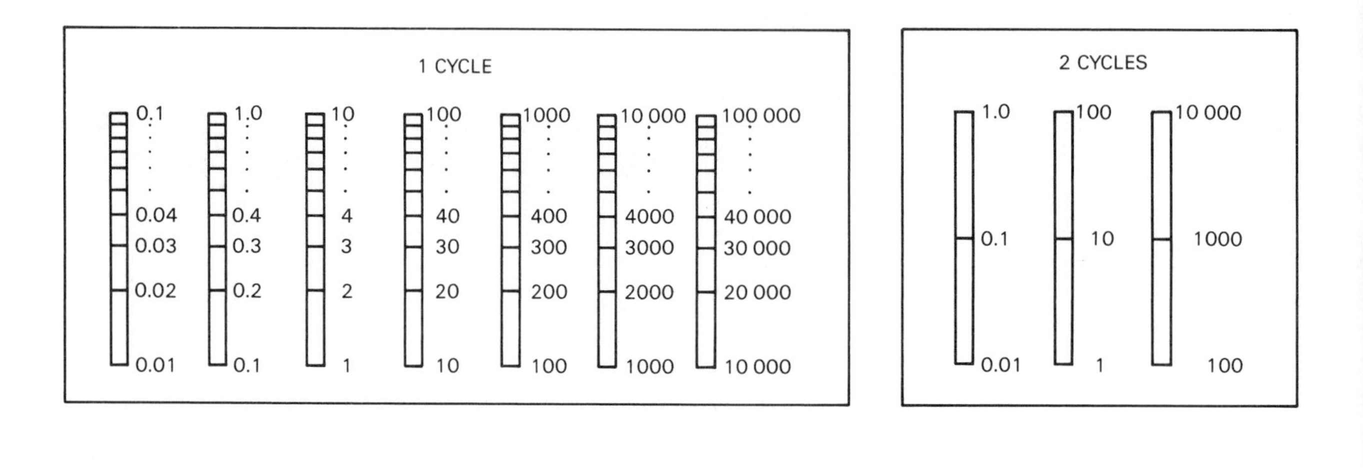

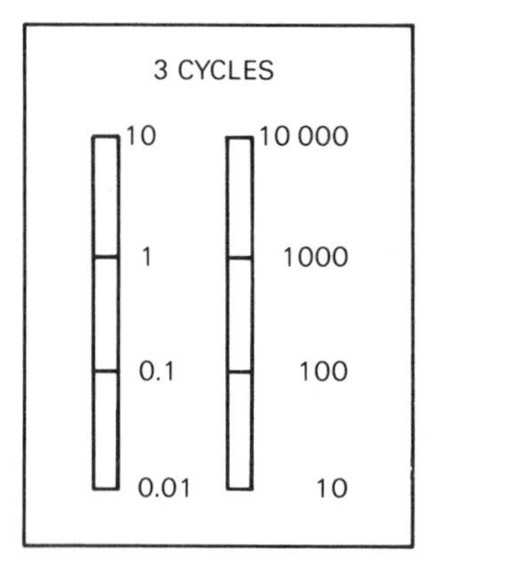

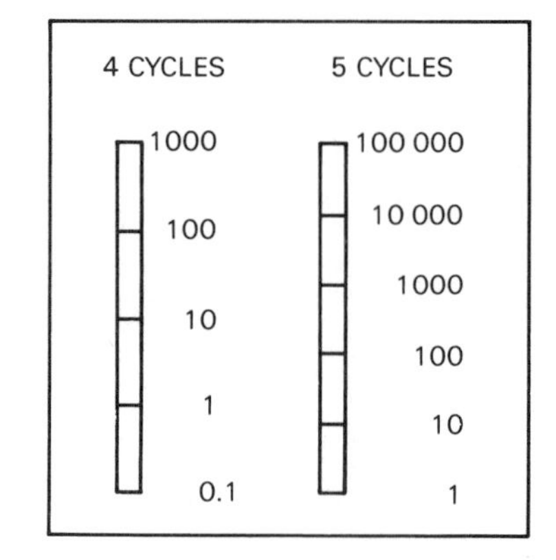

EXERCISE

The values of X and Y from an experiment were as shown in the table.

X	0.2	0.8	1.1	2.0	3.2
Y	0.16	10.2	26.6	160.0	680.0

1 How many cycles are required to accommodate the X values?

2 How many cycles to accommodate the Y values?

3 What type of log grid would you therefore use?

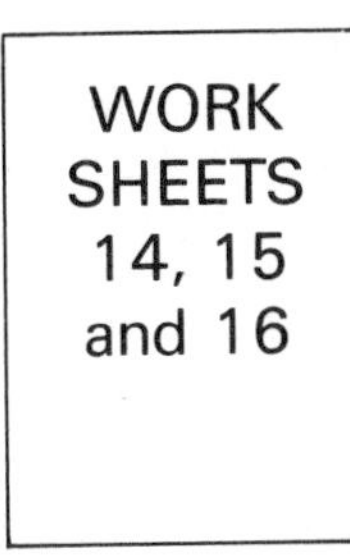

The log × linear graph

The advantage of using log grids

EXERCISE

In an experiment (Experiment 1) to study population growth in a species of yeast, a sample of cells was placed in a nutrient solution. At hourly intervals the flask was shaken and a sample taken and placed on a haemocytometer slide. Counts were taken in each of 5 squares and a mean value obtained. The results of this experiment are given on Work Sheet 14.

Haemocytometer: × 10 objective
Improved Naubauer
Rulings

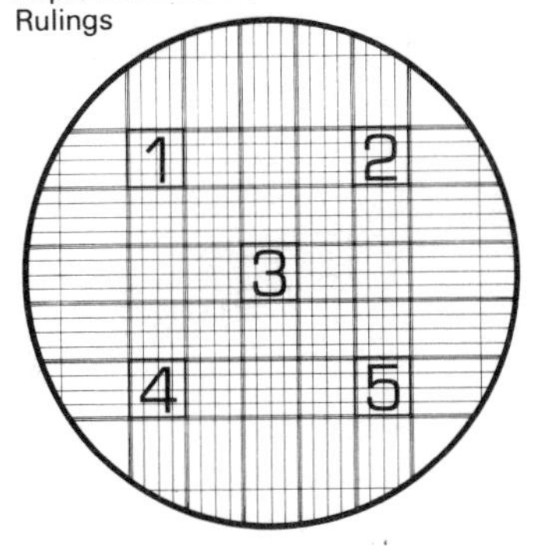

1 Draw a line graph of these results on the linear grid provided.

This shape of graph is often encountered in Biology. It is known as an ***exponential*** curve.

2 Extrapolate the line you have drawn to find the number of cells which would be present after 7 h and after 14 h.

3 Look up log tables or use your calculator to convert the number of cells into the log number of cells. Enter your answers in the table given on Work Sheet 15.

4 Plot, on the linear grid provided, the age of the culture against the log number of yeast cells.

5 What sort of line is produced?

6 Extrapolate this line to estimate the number of yeast cells at 7 h and at 14 h.

7 Plot the original results of age against the number of cells **directly** on to the log (two cycle) × linear grid (Work Sheet 15). The grid contains two cycles for reasons you will find out later.

8 What sort of graph do you get? Actually the graphs produced in questions 4 and 7 should be identical.

9 Check by extrapolation that the results you obtained in question 2 are the same.

10 From this exercise you should have learnt something about the advantages of using log grids. Suggest what these are.

The log × log graph

EXERCISE

In an experiment to determine the effect of changes in light intensity on the rate of oxygen production in *Elodea* using a modified 'Audus' apparatus, it was decided to use the distance of a lamp from the *Elodea* as a measure of the light intensity. The lamp was placed at varying distances from a photoelectric cell and meter readings were taken. The results are given on Work Sheet 16.

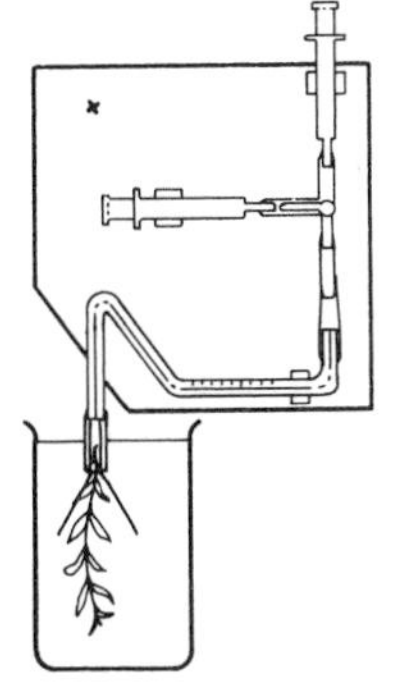

Photosynthometer ('Audus' apparatus)

1 Draw a graph of these results on the log two cycles × log one cycle grid provided (Work Sheet 16).

2.2.5 Scattergraphs (scatter diagrams)

Let us look at the concept of interdependence, attempting to visualise it by using graphs. There are really two questions concerning the relationship between interdependent variables to which we want answers:

(i) how intense is the relationship; and
(ii) what is the form of the relationship?

The intensity of the relationship is called the ***correlation***, and the form of the relationship is defined by the ***regression*** of one variable on another.

The situation here is quite different from that previously considered, where the *X* axis was occupied by the independent variable and the *Y* axis by the dependent variable. In this case we are considering an interdependent relationship. In some cases it might be possible to separate the variables into independent and dependent, e.g. if we were comparing stride length and leg length in humans it would be reasonable to suggest that leg length is the independent variable and stride length the dependent variable since the stride depends on the length of the leg. Leg length could thus be assigned to the *X* axis and stride length to the *Y* axis. In most cases of interdependence it is not possible to do this since the variables do not exhibit any dependence: a bean is not broader because it is longer or *vice versa*. To overcome this problem it is more convenient to label the axes *Y*1 and *Y*2 instead of *X* and *Y*; this tends to emphasise the interdependent nature of the relationship and eliminates difficulties.

Correlation

In order to visualise the intensity of a relationship between two variables we can make use of a particular kind of graph called a ***scattergraph***, or scatter diagram. The grid for a scattergraph is set out in the usual way, the two axes drawn at right angles to each other, but in this case they are labelled *Y*1 (across the page) and *Y*2 (up the page); the scales may be linear or logarithmic.

The data are given as one value for each of the two variables, e.g. length ***and*** breadth of particular privet leaves. There is therefore a number of values for both *Y*1 and *Y*2, but they are always given in ***pairs***. Each pair of values is separate and distinct.

Normally it is advisable not to use displaced origins since the proper origins may be required for analysis but this decision depends on the nature of the data and whether or not analysis is going to be carried out.

Each pair of values acts like the co-ordinates of a point on the grid so that each pair of values is represented by a dot. It is not necessary to put a circle around the dot since in most cases the line will not pass through the dots.

The points which correspond to each of the item pairs are located on the grid as below.

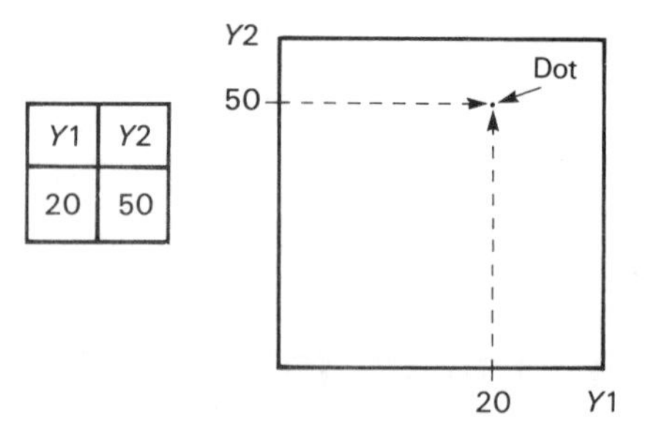

In scattergraphs there is no joining up of the points by a line since there is no continuity involved, each example being quite distinct from the rest. All we end up with is a scattered set of dots, hence the term scattergraph.

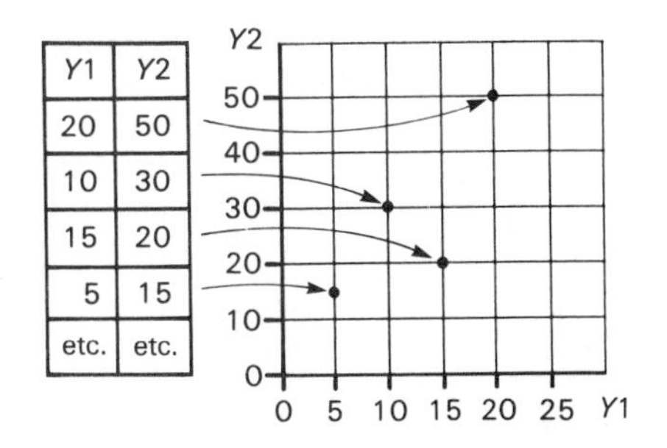

EXERCISE

1 On Work Sheet 17 there is a set of hypothetical data. On the grid provided plot a scattergraph.

Pictorial scattergraphs

On certain occasions we may want to compare scattergraphs of two different populations. In such cases the dots can easily be confused, particularly when they are interspersed with one another. To overcome this problem different coloured dots or even different symbols can be used.

EXERCISE

Two species of rush, the Soft Rush (*Juncus effusus*) and the Compact Rush (*Juncus conglomeratus*), are normally quite distinct from one another. *Juncus effusus* has pale flowers and a loose inflorescence; *Juncus conglomeratus* has dark flowers and a dense inflorescence.

Juncus effusus *Juncus conglomeratus*

The data given in the table below refers to the bract length and the number of stem ridges of typical specimens of the two species. Symbols can be used to represent the two species: ○ for pale flowers and ● for dark flowers; ∨ for loose inflorescence and | for dense inflorescence. *Juncus effusus* can thus be represented by the symbol ♉, and *Juncus conglomeratus* by ⫰.

No.	*J. effusus*		*J. conglomeratus*	
	Bract length (mm)	**Number of stem ridges**	**Bract length (mm)**	**Number of stem ridges**
1	155	58	55	18
2	248	53	65	19
3	240	49	75	20
4	200	47	95	18
5	155	46	120	19
6	105	43	30	18
7	135	41	70	18
8	145	41	110	17
9	175	43	125	16
10	90	35	35	15
11	110	37	115	17
12	120	35	95	15
13	125	33	45	15
14	145	36	75	14
15	155	33	50	15
16	180	38		
17	190	38		
18	175	33		
19	180	33		
20	205	25		
21	240	36		
22	245	32		

A scattergraph, using these symbols, was drawn from the data.

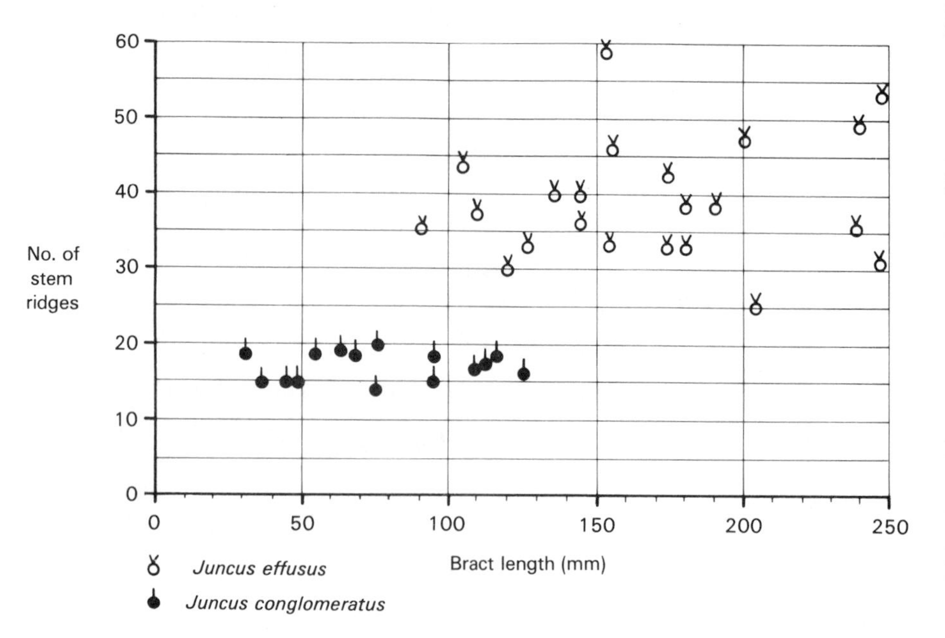

WORK SHEET 18

You will notice that the two populations are distinct from each other. Occasionally, however, one comes across specimens which are difficult to categorise into either species – they possess inflorescences which are intermediate between loose and dense. The table on Work Sheet 18 gives information on ten of these non-typical specimens.

1 Present the data as a scattergraph on the grid provided, so that it can be compared with the scattergraph on this page. The same symbols for pale or dark flowers and loose or dense inflorescence should be used. The intermediate inflorescence can be represented by the symbol ⩛.

2 Do the two scattergraphs give evidence to suggest that the two species may be hybridising? [S]

Three-dimensional scattergraphs (optional)

If we want to illustrate the correlation between three variables we can use the isometric grid to form a three-dimensional scattergraph. In many respects plotting such a graph is similar to plotting the three-dimensional line graph which was described in section 2.2.3 (p. 19).

Consider the following data on the length, breadth and mass of broad beans.

Bean number	Length (mm)	Breadth (mm)	Mass (g)
1	15	8	0.7
2	18	10	0.8
3	20	10	1.0
4	22	13	1.4
5	26	15	1.5
6	27	19	1.9

We can see from these measurements that the longer the bean, the wider and heavier it is, and *vice versa* (what we would expect), but what would a three-dimensional scattergraph look like?

EXERCISE

1 Draw the graph of the data on the isometric grid on Work Sheet 19, following the instructions, given here.

(i) Sort out the scales: length from, say, 10 to 30 mm, breadth from 5 to 20 mm and mass from 0 to 2.0 g.

(ii) The main 'box' can now be drawn on the grid and the scales inserted. (We will try moving the scales around later to find out what effect this has on the appearance of the graph.)

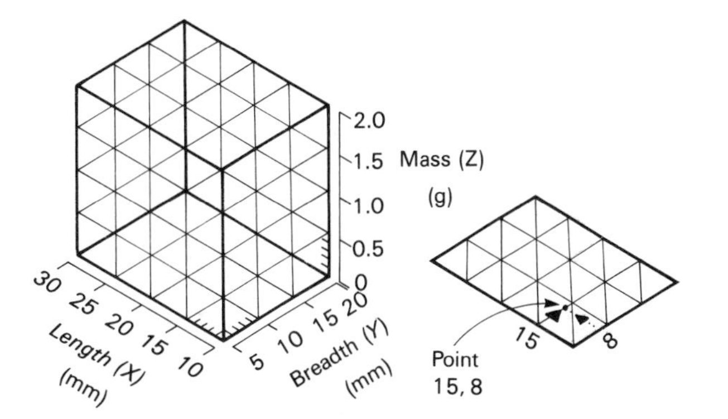

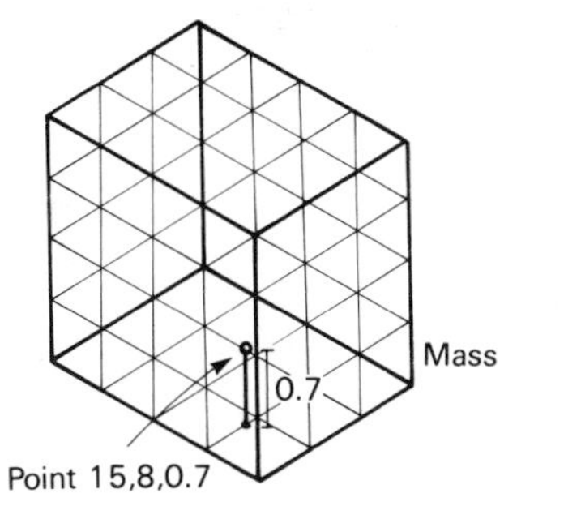

(iii) As in the three-dimensional graph, consider the base of the 'box' as an ordinary linear grid. In this case we are dealing with the first two values, e.g. bean 1: length 15 mm, breadth 8 mm. At the intersection of these two lines, a large dot is placed.

(iv) The third value for bean 1 was a mass of 0.7 g. Count vertically seven divisions from the dot you have drawn in (iii), draw a small circle at this point, and then join it to the dot with a solid line, so that is looks like a pole sitting in the middle of the box.

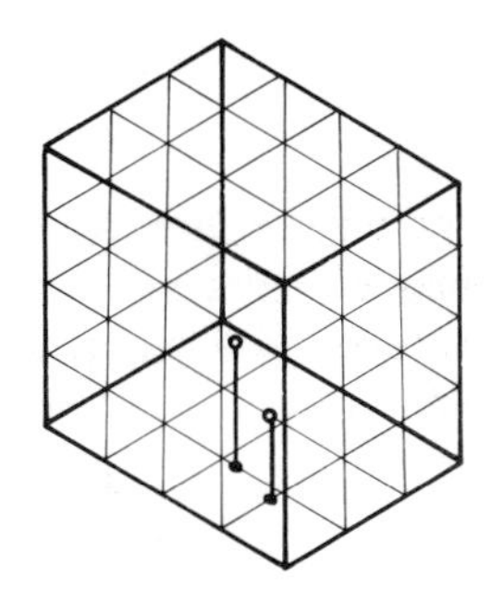

(v) Bean number 2 is 18 mm wide and has a mass of 0.8 g. Proceed as in (iii) and (iv) to find and plot its position. Your graph should now look as on the left.

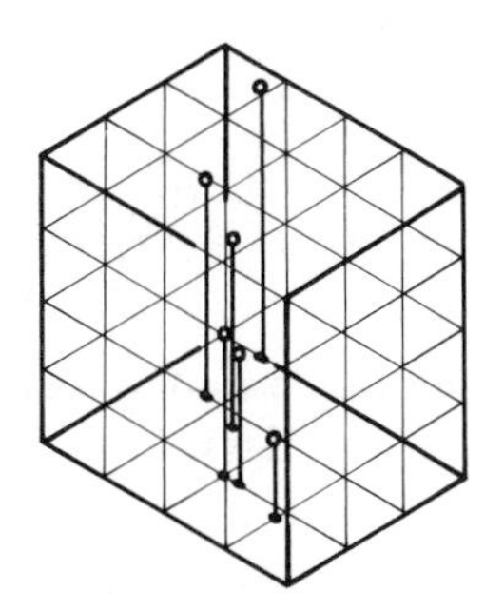

(vi) Continue plotting until all six beans are accounted for. Your final graph should look something like that on the left.

We can see that the points lie close together linearly, but only with regard to the relationship between length and breadth. For the third variable, mass, although the points get higher the further back they are placed, the relationship is more difficult to visualise.

We can overcome this problem by reversing one of the scales, e.g. length.

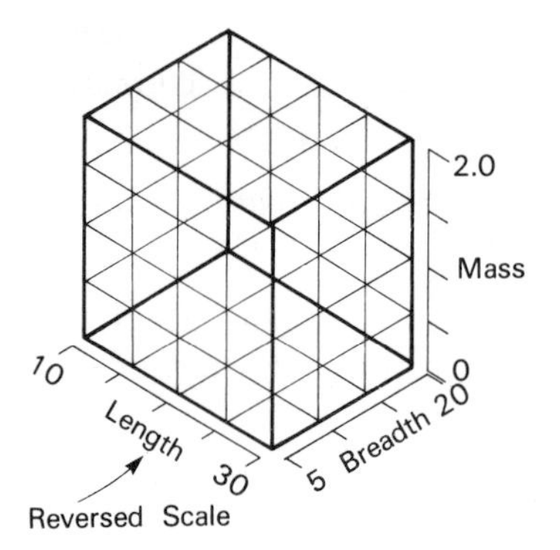

2 Plot the scattergraph again, but on the second grid (Work Sheet 19) with the scales entered as on the left.

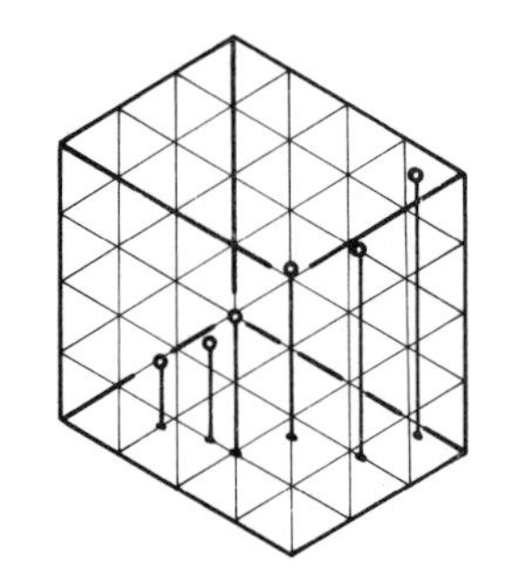

Your graph will now look something like that on the left. The relationship between all three variables is now much easier to visualise.

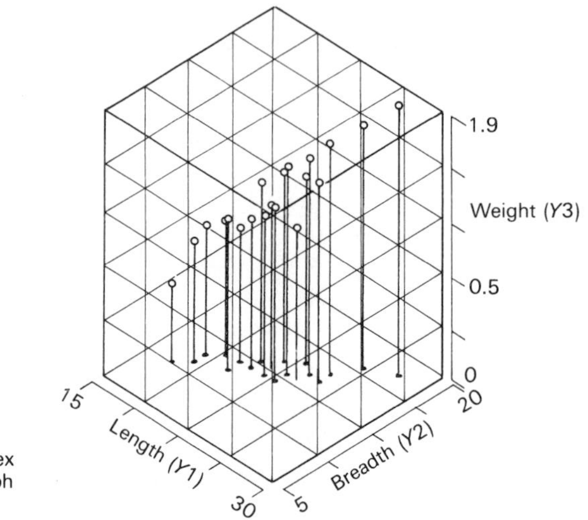

A more complex 3D scattergraph

2.2.6 Frequency distribution graphs

You saw in section 1.3.3 that data collected in Biology often involves the number or frequency of observations falling within particular categories or classes. Such frequency distributions lend themselves readily to graphical techniques, conveying the information in a simple, rapid and concise way.

In most graphs of frequency distribution the categories or classes are placed on the X axis, and frequencies are measured along the Y axis. There are exceptions to this which we will look at shortly. To avoid any confusion with line graphs it is customary to label the Y axis the f axis (for frequency). It is usual to begin the f axis at zero, but the X axis can have a displaced origin if necessary, to make best use of the grid.

Point diagram

EXERCISE

The simplest type of frequency graph is the point diagram, where dots are placed at points which are defined by the mid-points of the classes (X axis) and the f values.

1 Plot a point diagram of the *Patina* data from section 2.1.3, (p. 9) on the grid provided on Work Sheet 20.

The point diagram is not really satisfactory since the scattered dots don't really appear to form a pattern. A graph like this can also be confused with the scattergraph (see section 2.2.5, p. 26) which is totally different from a frequency distribution.

Frequency polygon

If the dots of the point diagram (Work Sheet 20) are joined together by straight lines, the result is a frequency polygon.

2 Join up the points on the point diagram and you will notice that the visual pattern is much clearer. Such graphs are, however, not often used except in special cases.

WORK SHEET 21

The data given on Work Sheet 21 are drawn from random samples of a fish species trawled from 1945 to 1948 (during the Second World War very little fishing was carried out). The frequency is related to the length of the fish.

1 Plot points for these data on the grid supplied (Work Sheet 21), and then join up the points with straight lines to form a frequency polygon.

2 The resulting four 'peaks' stand out clearly on this frequency polygon. What do these peaks represent?

The number of classes and thus the class interval is important. If there are too many classes, the detail will obscure any pattern that might be present, but equally if there are too few classes, the pattern can be smoothed over.

up to here

WORK SHEET 22

Look again at the *Patina* data (Work Sheet 20). There were ten classes and the class interval was 1.2 mm. If the class interval is doubled to 2.4 mm the number of classes will be halved from ten to five (see table on Work Sheet 22). What effect will this have on the pattern of the distribution?

1 Draw a frequency polygon of these data on the grid provided (Work Sheet 22), and compare it with the one you drew of the ten classes (Work Sheet 20).

2 Are any details obscured?

Bar graphs; bar charts

The bar graph should be used for data which is qualitative or discrete, i.e. where the categories are clearly separate and distinct from one another. The main characteristic of the bar graph is that it consists of bars which are separate from one another, and the ***height*** of the bars is proportional to the frequency. The bars can be of any convenient width, and placed in any convenient order. The separate nature of the bars helps to emphasise the discrete nature of the categories.

Line bar graph

In this case the bars are so narrow that they are simply lines. This makes the graph very easy and quick to construct.

EXERCISE

Pitfall traps were set in a sheep farming area, and every month for a year the mass of ground beetles caught in them was measured. The results are given on Work Sheet 23. The months are the categories and so are placed along the X axis. These are naturally separate and independent entities. The mass of ground beetles in this case represents the frequency and so is measured along the f axis. The mass of beetles actually represents the numbers of beetles. The values of f for each month are represented by lines drawn from the X axis.

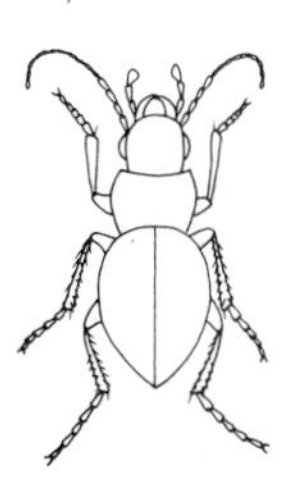

Carabid beetle

1 Draw a line bar graph on Work Sheet 23.

2 These data could equally well be represented by a frequency polygon. Draw a frequency polygon on the grid on the same Work Sheet and compare it visually with the line bar graph.

Narrow bars

The more traditional type of bar graph consists of narrow bars instead of lines representing the frequency. The tops of the bars not only trace visually the rise and fall of the frequencies, but the bars also tend to accentuate the actual quantities involved. The line graph only draws our attention to variations in the frequencies.

WORK SHEET 24

EXERCISE

Shoots of the Canadian pond weed *Elodea* were placed under various coloured filters, and the time taken for the release of twenty bubbles from the cut end of the stem was recorded. The results are given on Work Sheet 24.

1 Construct a bar graph of the data.

Shading the bars helps to accentuate them, drawing more attention to the quantities involved. Always try to shade as neatly as possible. Alternatively, since the categories in this case are various colours, you could colour the bars, eliminating the need to label them.

It is always a good rule to follow that ***all*** lettering and numbering on a graph should be across the page so that the reader doesn't have to turn the graph on its side to read it. You will notice that in the last graph the labels for the narrow bars had to be written sideways in order to fit them properly. One possible solution to this problem would be to code the bars with different types of shading and supply a key. This can, however, be confusing, particularly if there are quite a few categories.

Often it is better to ignore the general rule that the X axis runs along the base of the grid, and turn the graph on its side so that the *X* axis runs down the left-hand side. The bars will then lie across the graph and the categories can be labelled with the labelling the right way up; the labels can even be placed inside the bars.

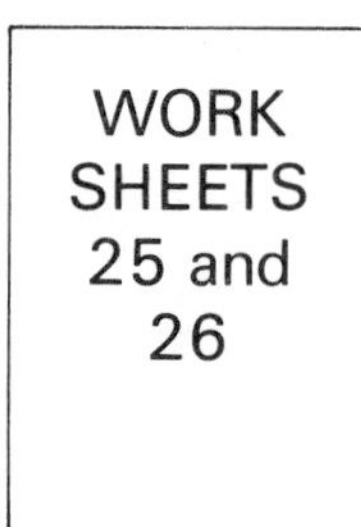

EXERCISE

The populations of a number of invertebrates were determined from an area of chalk grassland. The results are given on Work Sheet 25.

1 Plot the data on the grid provided. In this case the bars should run across the grid. Write the labels inside the bars.

Occasionally, you may come across data which have a very large range, the scale going from very small values to very large ones. You should already know that log scales can help to alleviate this problem. Remember, however, that log scales visually minimise the differences between the categories.

Hummingbird

EXERCISE

The data on Work Sheet 26 give the frequency of wing beats (per second) of various animals.

1 Plot this information as a bar graph on the log × linear grid provided (Work Sheet 26).

Compound bar graphs

As in the case of the line graph, a number of different groups of categories can be plotted on the same grid so that they can be compared visually. There are various possibilities, depending on the available data.

Two sets of data; one *f* scale

Often in an investigation two sets of data are obtained using similar techniques from different locations, or situations.

EXERCISE

On Work Sheet 7 data was given of the population of invertebrates from an area of chalk grassland. Further data was also obtained from an area of heathland using the same sampling and trapping methods. The results are given on Work Sheet 27.

The results for the two areas, chalk grassland and heathland, have to be kept distinct from each other by the use of different shading of the bars, or by the use of different colours. The bars are placed up the grid this time, with the X axis along the base, since with the double bars there is room for the labels.

1 Complete the bar graph on the grid provided (Work Sheet 27).

Lithobius forficatus

Two sets of data; two *f* scales

The two sets of collected data might be based on totally different units but we might still wish to compare them, so two *f* scales have to be employed. This is a similar situation to that in which two Y scales were used in line graphs. One *f* scale occupies the left-hand side of the grid and the other *f* scale the right-hand side.

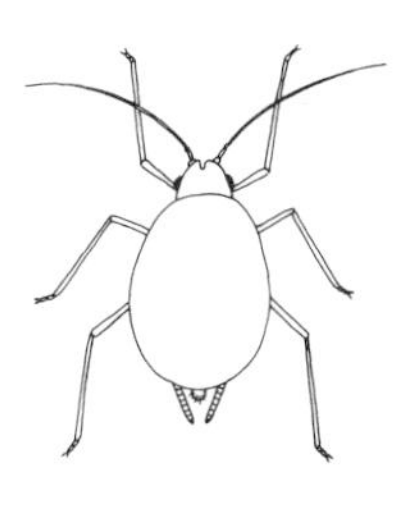

An aphid

EXERCISE

Certain aphids feed on the sugar beet plant. One in particular, the Green Peach Aphid, *Myzus persicae*, feeds by probing several times, feeding for a few hours and then moving on to another plant. During the years 1940 to 1947, sugar beet plants were selected at random and the number of *Myzus* per 100 plants counted. Sugar beet plants can become infected with beet yellow virus, with serious economic consequences. A hypothesis was put forward that this virus was transmitted by the Green Peach Aphid. For the same years, 1940 to 1947, the percentage infection of sugar beet plants with this virus was also measured. The results are given on Work Sheet 28.

1 Construct a bar graph of these data on the grid provided (Work Sheet 28), using the two *f* scales shown. You will notice that the *f* scale for the number of *Myzus* is logarithmic, since the range of values is so large. Where there are two grids, as in this case, it would be very confusing to have the lines included. Obtain the positions of the top of the bars by setting a ruler across the grid.

2 Does the graph support the hypothesis that *Myzus* is the possible vector for beet yellow virus?

Now try this example of two *f* scales, but this time place the bars sideways on the grid. Again we are looking for a possible relationship between the two variables.

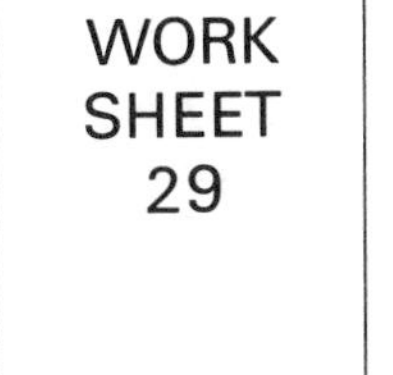

EXERCISE

The death rates from coronary heart disease per 100 000 people from different countries are given on Work Sheet 29, together with the sugar consumption per person per year for the same countries.

1 Construct a bar graph of these data on the grid (Work Sheet 29).

2 Is any relationship between the two variables evident from the graph?

Three sets of data

When there are three sets of data the visual appearance of the bar graph becomes more complex, but the trends of each group are still clear enough to be distinguished. Various types of shading can be used to good effect. It is necessary, of course, to put a key under the grid to identify the various bars.

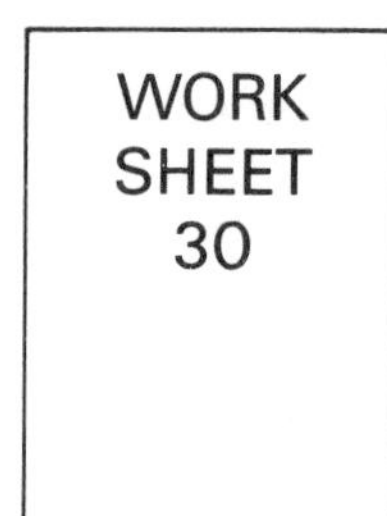

EXERCISE

The stomach contents of trout of various sizes were examined to obtain information on their diet. The results, given on Work Sheet 30, summarise the findings, the data being grouped according to the size of the trout. The figures for the different organisms show the percentage by volume of the organisms in the diet.

1 Draw a three-group bar graph on the log × normal grid provided (Work Sheet 30). Place the bars so that they go up the grid. There should be enough room for labelling of the *X* axis. Draw the graph as neatly as possible.

More than three sets of data

When there are four or more sets of data the graph becomes increasingly difficult to analyse visually. It becomes virtually impossible to ensure that any one group can be clearly identified from the others.

Example

Plankton were taken from the North Atlantic each month for a year and samples analysed for the numbers of five different species of copepod. The results are shown below.

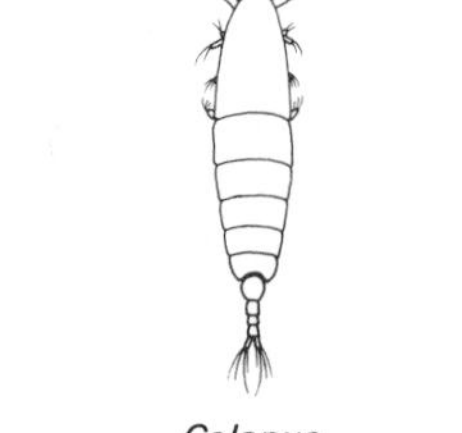

Calanus
(a planktonic copepod)

Month	Species				
	A	B	C	D	E
Jan.	0	110	40	50	0
Feb.	0	100	45	55	0
Mar.	80	95	120	105	0
Apr.	100	70	140	100	90
May	110	80	200	110	95
Jun.	20	30	150	100	100
Jul.	100	80	110	110	100
Aug	80	300	120	0	140
Sep.	70	400	130	0	200
Oct.	230	300	140	0	220
Nov.	80	70	90	0	150
Dec.	0	80	50	0	90

[S]

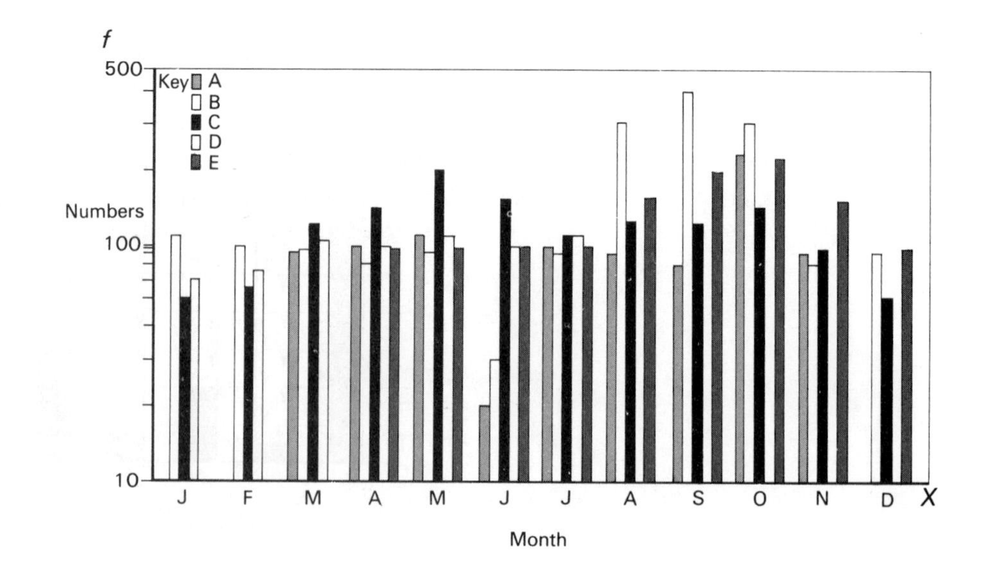

When there are as many as five sets of data, as in this case, the graph becomes visually confusing, so it is often better to use frequency polygons, one for each set of data, set above each other for comparison. That for species C is shown below.

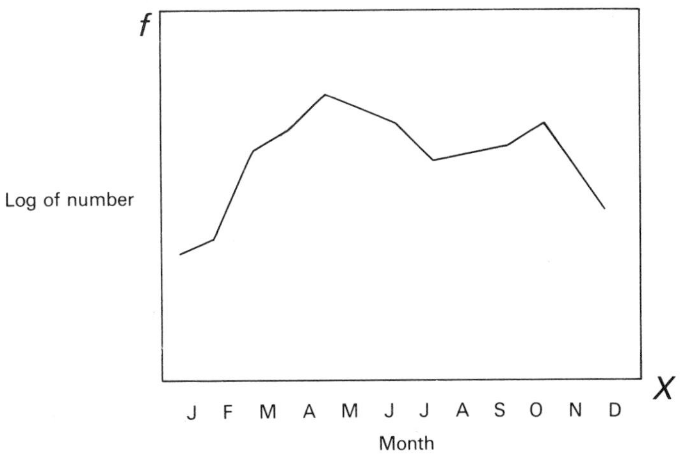

The percentage bar graph

A variation of the compound bar graph is the percentage bar graph, where the complete bar is divided into the relative percentages of the components, which therefore add up to 100.

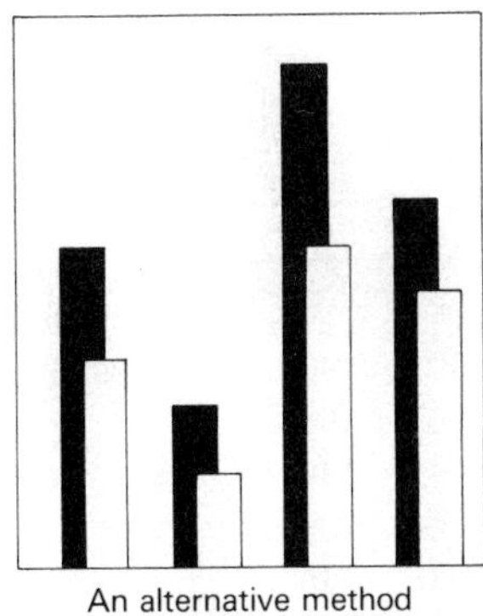

An alternative method of drawing bar graphs

Example

Various sampling techniques were used to estimate the percentages of a number of different invertebrate groups present on the ground (pit-fall traps), on vegetation (by searching and sweeping) and in the air (water and sticky traps). The results and bar graph are given below. Note that the bars go up the page since the data represent a vertical distribution from the ground to the air. The categories are labelled at an angle to make them easier to read.

	Percentages		
	On the ground	**On vegetation**	**In the air**
Araneida (spiders)	67	33	0
Hemiptera (bugs)	4	54	42
Diptera (flies)	20	37	43
Coleoptera (beetles)	7	52	41
Hymenoptera (bees, wasps)	30	26	44

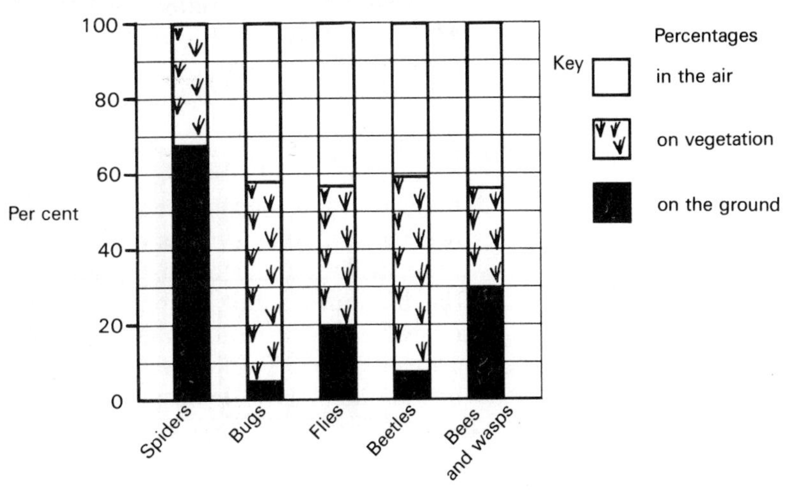

Histograms (histographs; block graphs)

The histogram should be used when the X variable consists of a ***continuous*** variable (see section 1.2.3, p. 3) and the data have been arranged into classes. The f variable is the frequency falling within each particular class.

Simple histograms

In section 2.1.3 you saw how the data on the length of *Patina* was arranged into classes. A histogram of the arranged data looks like that below.

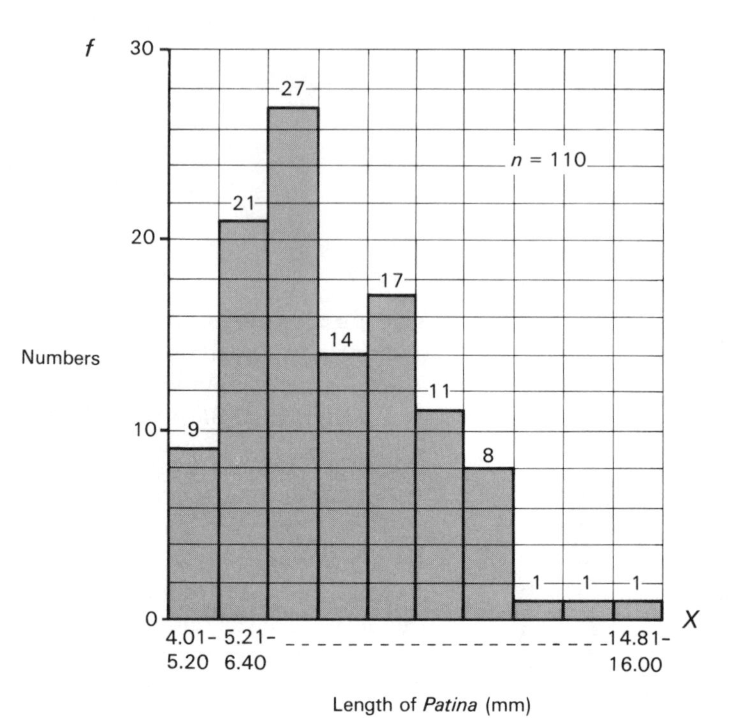

Note the following points about the histogram. The classes are placed along the X axis, the limits of each class being clearly shown, and each class is mutually exclusive, i.e. the boundaries are clear. (Sometimes only the boundary points or the mid-points are given.) The numbers of *Patina* are measured along the f axis. Lines are drawn up from the X axis at the limits of each class and the class columns are topped by lines which indicate the number of *Patina* falling into each class. There are therefore no gaps between the columns. At the top of each column it is useful to show the frequency which the column represents, and also to indicate the total number of organisms measured. The axes are appropriately labelled. Finally the columns are shaded to make the distribution stand out.

The whole ***area*** shaded represents the total sample studied. Since the columns are equal in width, the heights of the columns are proportional to the frequency, but always remember, it is the ***area*** of the column which represents the frequency. It is very rare that columns of different widths would be used. The number of classes can have an important bearing on the shape of the distribution and thus on the conclusions that can be drawn from it. The following two histograms show the same *Patina* data but with five classes and sixteen classes. The vertical lines have been omitted and only the external boundaries of the distribution are drawn. This tends to accentuate the nature of the distribution. (See also section 2.2.6, p. 30, and Work Sheets 20 and 22.)

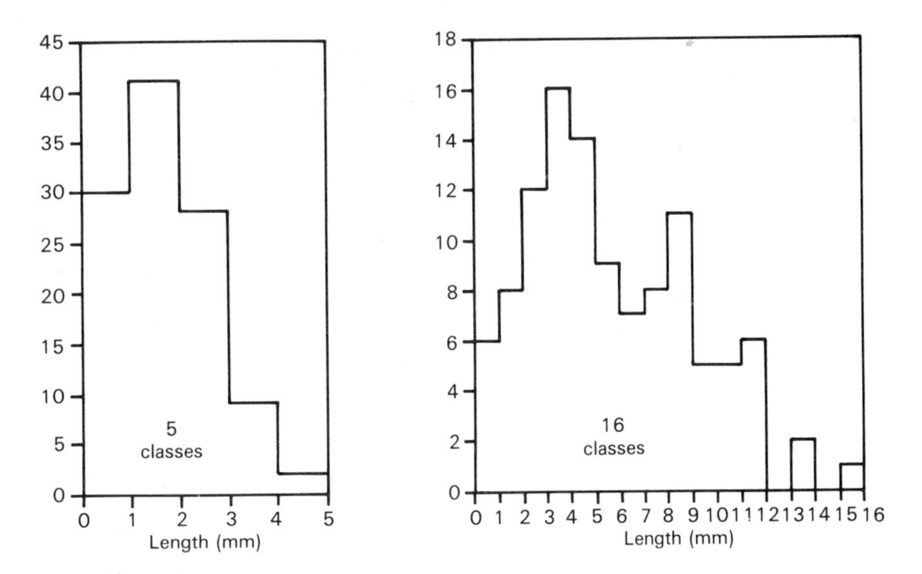

WORK SHEET 31

EXERCISE

One hundred and twenty-five privet leaves were removed at random from the top of a high hedge and their lengths were measured carefully to the nearest millimetre. The results are given on Work Sheet 31.

1 Arrange the data into nine classes, using the tally chart, and draw a histogram on the grid provided.

Compound histograms

In some investigations comparison between different groups of data may be required. These can be contained on one grid. It is not, however, a simple matter in such cases to follow the trends of all the components. Really the only component which can be followed with ease is the one at the bottom, the base of which is a straight line. The rest have 'floating' bases which bob up and down at different rates to various levels depending on the values of the variable below.

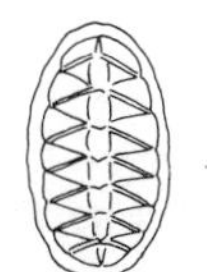
Lepidochitona cinereus

Example

In an experiment to investigate behaviour in the chiton (coat-of-mail shell), a mollusc found on the sea-shore, groups of them were subjected to the following conditions:

Group A – overhead illumination from heat-filtered daylight.
Group B – overhead illumination from a bench-lamp.
Group C – total darkness.

The number of chitons moving a certain distance over 10 min was recorded. The results are given in the following table (mid-points of classes shown).

Distance moved in 10 min mm	Number of chitons		
	Group A	Group B	Group C
0	–	–	14
30	1	5	2
60	3	11	1
90	7	4	2
120	4	–	1
150	2	–	–
180	2	–	–
210	1	–	–
240	–	–	–

[S]

These data can be drawn as a composite histogram as below. Only the trends of group A can be followed clearly, the trends of the other groups being too confused to follow.

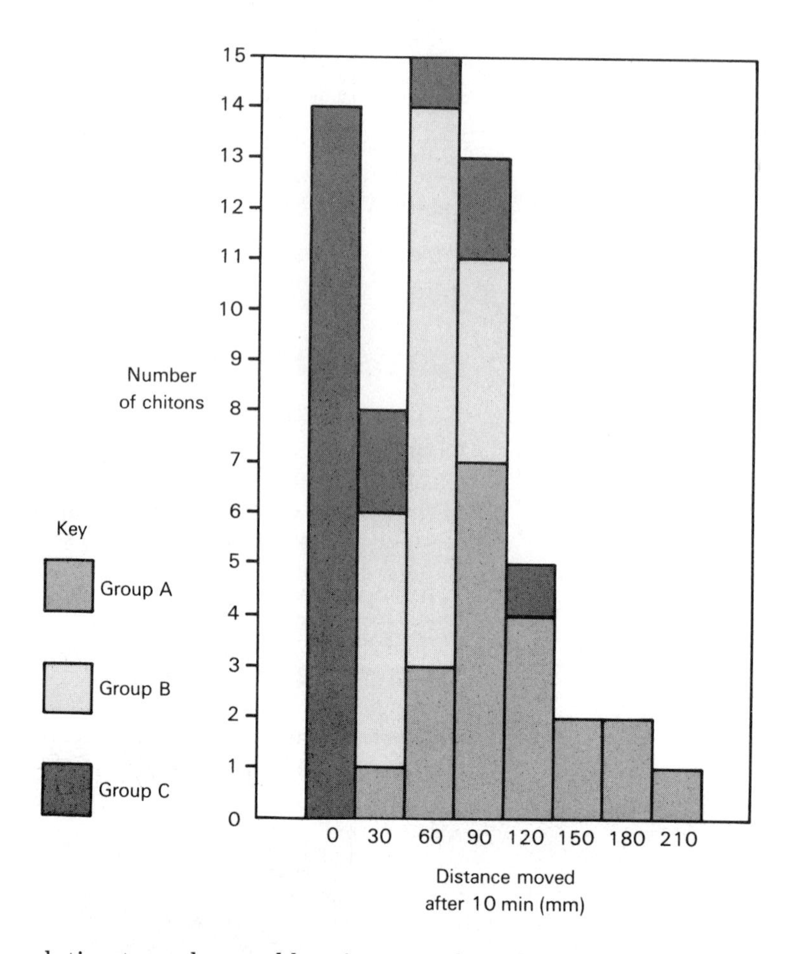

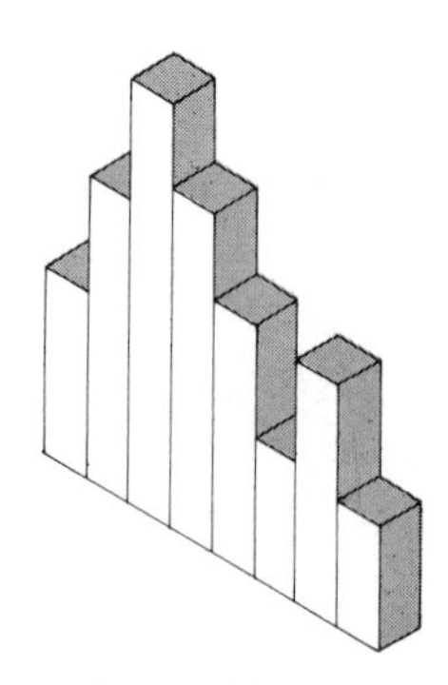
A stereohistogram (often used in computer graphics)

The solution to such a problem is to graph each of the groups separately, placing them directly above one another in a cumulative fashion. Each group can now be seen clearly and can readily be compared with one another.

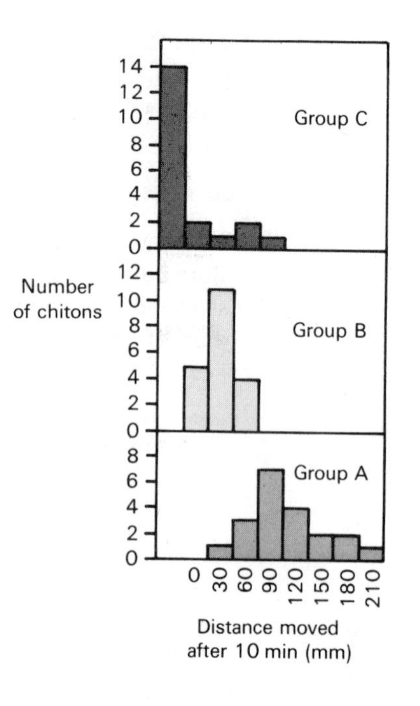

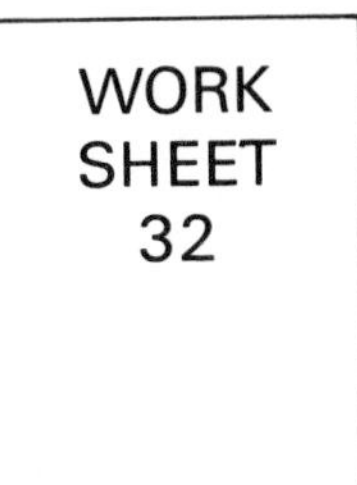

EXERCISE

Littorina saxatilis, a periwinkle found on the sea-shore, exhibits such a variety of shell patterns and reproductive habits that taxonomists have had considerable difficulty in classifying the group. Some workers have described four sub-species – *rudis, tenebrosa, jugosa* and *neglecta*. More recent work tends to place these into species. This problem does not have any important bearing on the exercise.

Fifty of each sub-species were collected at random on one day from a rocky shore and the shells were weighed. A frequency table was drawn up from the results (Work Sheet 32).

1 Plot these results as histograms so that the distributions of each sub-species may be compared.

Littorina saxatilis

Pictographs

Pictures can often be used in the design of graphs to enhance their presentation by registering a meaningful impression in our mind, even before we think carefully about them. Some graphs can be almost purely pictorial. If, for example, we wanted to show the distribution of seaweeds on a particular seashore we could represent it as below. In this case the symbols actually look like the seaweeds.

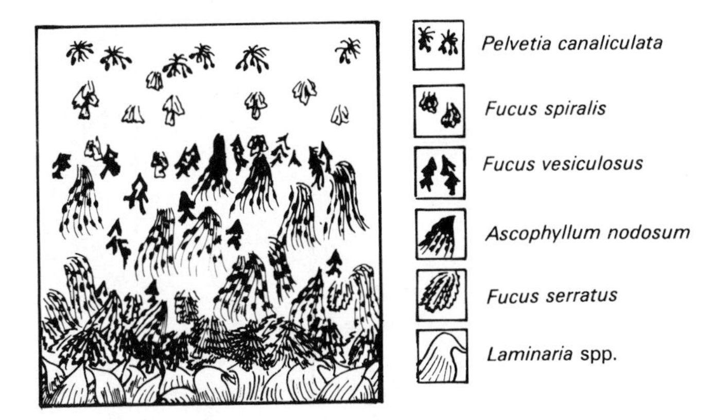

This type of graph gives no indication of the quantitative nature of the distribution. Other methods will be described later which overcome this problem. This type of graph can be very effective in particular circumstances.

Graphs and some form of pictorial representation can be mixed for any data for which this treatment would be suitable.

Kite diagrams

In the previous section you saw how the distribution of seaweeds on a seashore could be represented. If we wanted to quantify the distribution of the seaweeds we could devise a scale of abundance that was related to the amount of the surface covered by the seaweeds, for example along a transect.

Abundance scale	Coverage
1	$<\frac{1}{20}$
2	$\frac{1}{20}-\frac{1}{4}$
3	$\frac{1}{4}-\frac{1}{2}$
4	$\frac{1}{2}-\frac{3}{4}$
5	$>\frac{3}{4}$

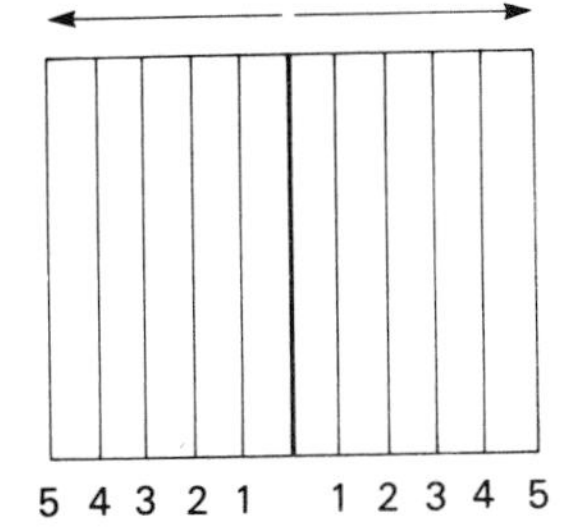

The X axis of the grid on which the graph is drawn represents the abundance and the Y axis the distance from low water.

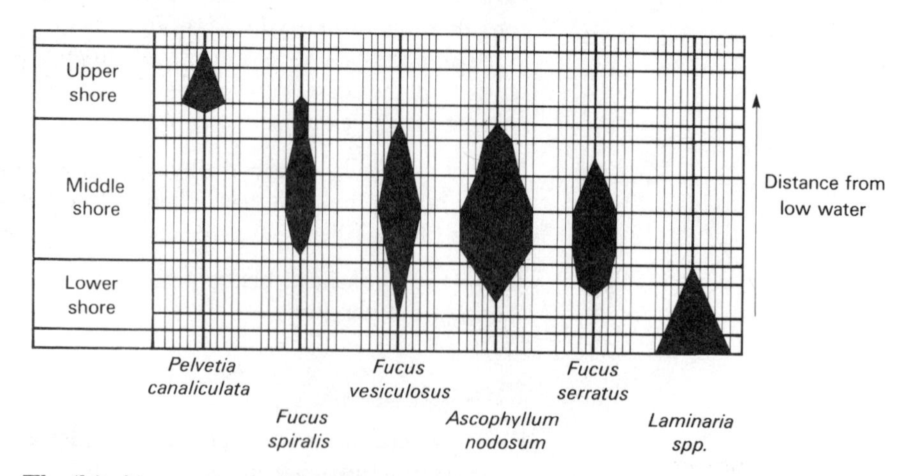

The 'kite' is constructed by plotting points on either side of a line according to the abundance scale. These are then joined together, 'tailed off' and finally shaded in. An example with values of 2, 3 and 1 is shown below.

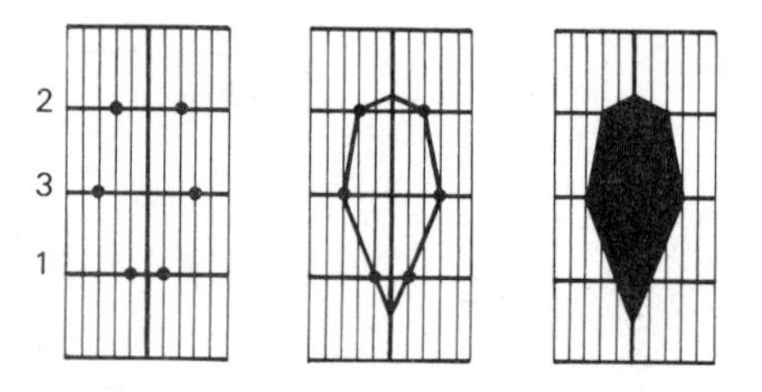

WORK SHEET 33

EXERCISE

A transect was laid down on a shore from the splash zone to the low water mark (LWM). At points along the transect, measured in metres from the low water mark, the animals present were scored according to the following abundance scale.

Description	Abundance scale
Rare	1
Occasional	2
Frequent	3
Common	4
Abundant	5

The results are given on Work Sheet 33.

1 On the grid provided draw kite diagrams to show the distribution of each of the animals.

2 Is there any evidence of zonation on this shore?

Ladder transect graphs

Another method of illustrating quantitative data of the distribution of organisms along a transect is the ladder transect graph.

The abundance scale given in the previous section can be illustrated by means of a bar graph as below.

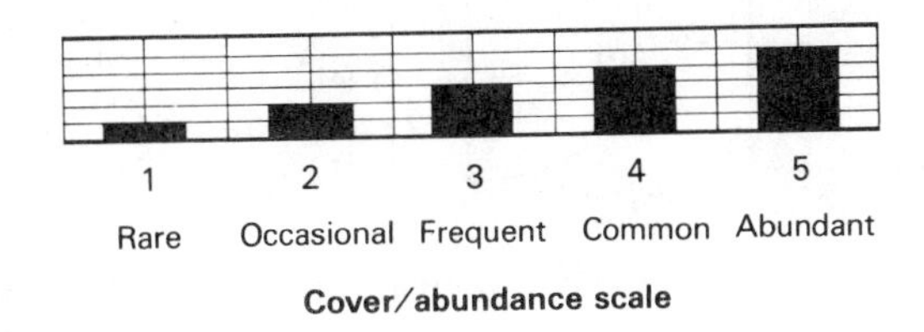

The completed graph for various seaweeds on a seashore would be as follows.

Splash zone
Upper shore
Middle shore
Lower shore
Sub-littoral

Pelvetia canaliculata
Fucus spiralis
Fucus vesiculosus
Fucus serratus
Ascophyllum nodosum
Laminaria spp.

WORK SHEET 34

EXERCISE

The data given on Work Sheet 34 shows the distribution of five plants across a swamp marsh and pond. The plants have been quantified according to the abundance scale given on p. 40.

1 Draw a ladder transect graph to illustrate the data.

2 Is there any evidence of zonation?

2.2.7 Circular graphs

Polar co-ordinate grid

This type of grid is circular. It is made up of one set of lines consisting of concentric circles about a point, and another set of radial lines. The point in the centre of the grid is known as the 'pole point' or more commonly just the 'pole'. The circles are determined by their radii, e.g. 5 mm, 10 mm, etc., and the radial lines are determined by their angular positions from a particular radius. Often the angle between radii is determined by the number of divisions required, e.g. the circle may be divided into 12 or 24 sectors, equivalent to the months of the year or hours in a day. Occasionally a large area in the centre is left clear since the lines near the centre may not be required.

There are two main ways in which we can use polar co-ordinate grids.

(i) The circle can be thought of as totality, which can then be subdivided according to the proportions of the constituent parts.

(ii) The usual linear grid is, as it were, bent around so that its two outer edges meet, on one plane.

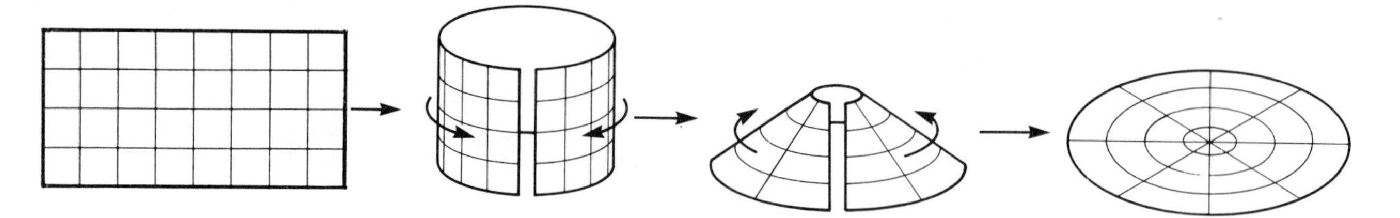

The usual rise and fall of the linear grid is thus replaced by a line which moves away from or towards the centre of the circle. Some practice is required before the information contained in this type of graph can be fully appreciated.

Pie charts (pie graph; sector chart)

This type of graph is a very striking and appropriate method of illustrating the proportions of an identifiable whole. The circle of the graph is divided into segments (presumably like pieces of a pie and nothing to do with pi and its relationship to a circle) which indicate the relative proportions of the parts.

How is such a graph constructed? The respective proportions, which may or may not be percentages, determine the angles of the sectors, the total circle being divided according to these angles. To find the angle of each sector the following simple formula is used:

$$d = \frac{V \times 360}{T}$$

where d is the angle of the sector in degrees;
V is the actual amount of the variable;
T is the grand total.

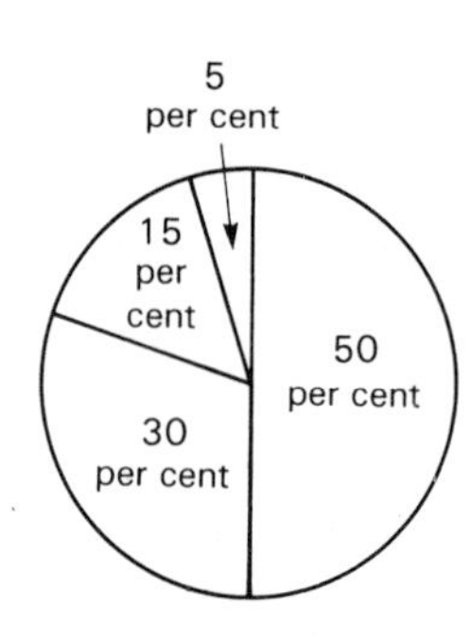

It is usual to arrange the sectors in rank order in a clockwise direction, starting at '12 o'clock', as can be seen in the pie chart on the left.

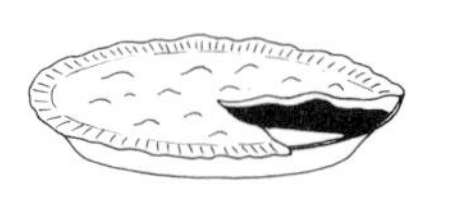

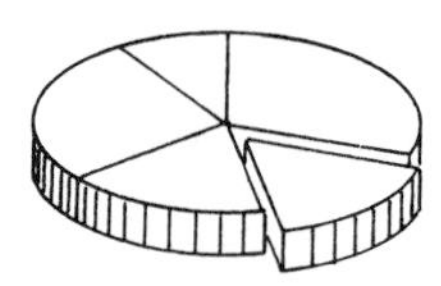

Example

The blood groups of a sample of 100 000 people living in the British Isles were determined. The results are shown.

Blood group	Number (000)
O	51
A	36
B	10
AB	3

From these results the values of d are calculated for each of the blood groups:

$$\text{Group O:}\quad d = \frac{V \times 360}{T} = \frac{51 \times 360}{100} = 183.6°$$

Group A works out at 129.6°, B at 36° and AB at 10.8°.

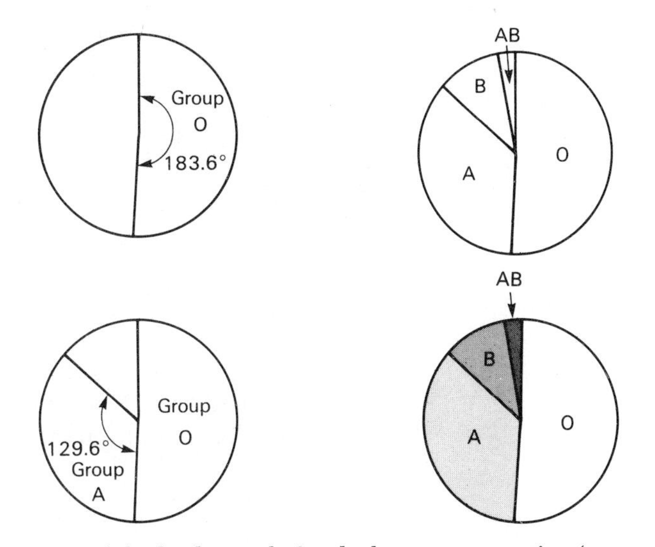

Starting at 12 o'clock, the angle for the largest proportion (group O), and thus the largest segment, is 183.6 °. This is drawn and the segment labelled. The next segment, in rank order, is group A, with an angle of 129.6 °. This is drawn and labelled. This process is continued until the pie chart is completed. It can then be shaded appropriately.

EXERCISE

The data on Work Sheet 35 shows the numbers of different orders of insects that were caught in two types of trap – water traps and sticky traps.

1 Using the grids provided, draw two pie charts of these data. The grids have been subdivided into 10 ° sectors to make your task easier.

Cyclical graphs

Graphs drawn on polar co-ordinate grids can be superior to other types of graphs when dealing with data that is cyclic, e.g. daily, monthly, yearly, since if such data is plotted on the usual grid the left and right hand edges of the graph are discontinuous whereas in reality they are continuous.

Example

The mean diameter of the largest ovarian follicle in the wood-pigeon was measured monthly throughout the year. A follicle diameter greater than 5 mm indicates an ovary which is close to egg laying. The results are shown.

Month	Mean diameter of follicle (mm)
Jan.	1.25
Feb.	1.50
Mar.	2.75
Apr.	3.75
May	4.00
Jun.	5.75
Jul.	6.50
Aug.	6.50
Sep.	5.00
Oct.	1.80
Nov.	1.50
Dec.	1.25

When these data are illustrated by a bar graph, the result is as on the next page.

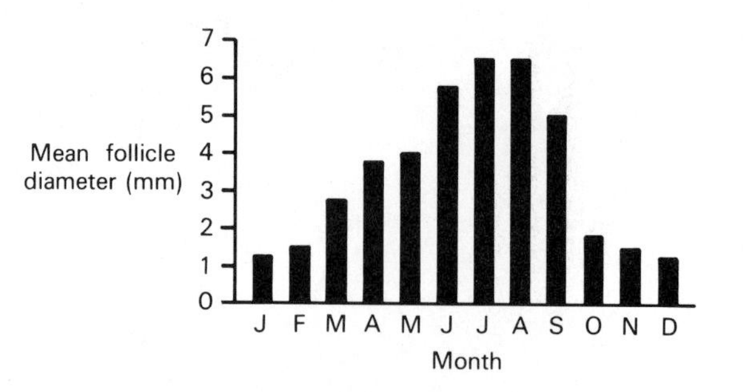

The same data have been drawn on a polar co-ordinate grid. The radii represent the months and distance along the radii represents the follicle diameter. Points are plotted on the radii at particular distances from the pole for each of the months. The points are then joined by straight lines. The result is a 'cyclical polygon'.

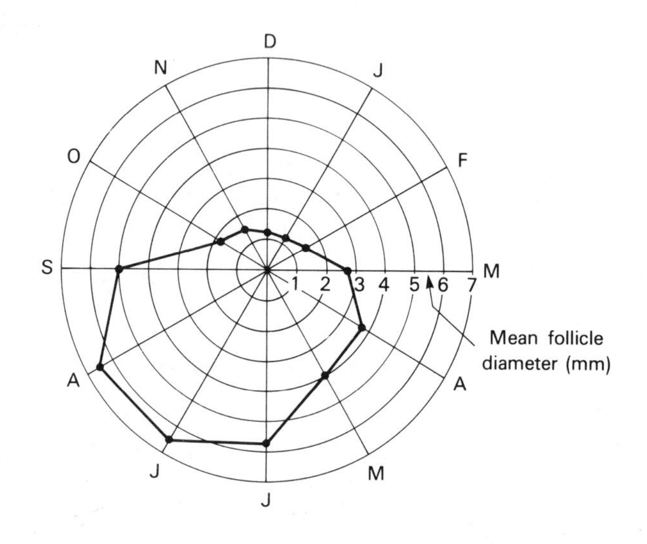

Rose diagram

As an alternative to the cyclical polygon, and in some ways more visually appealing, bar graphs can be drawn for each of the months. To prevent confusion of the bars at the centre, it is convenient to leave a 'hole'. Such a graph is often known as a rose diagram.

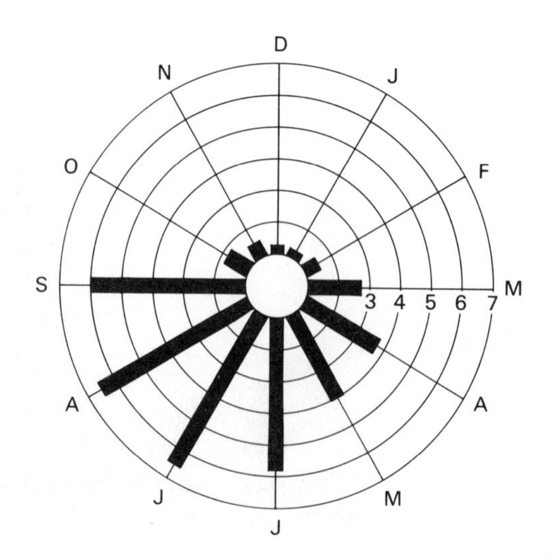

When dealing with more than one set of data, both the cyclical polygon and the cyclical bar graph can be used (in a similar way to the linear grid).

Example
Consider the following data, which gives the CO_2 output from decaying leaf litter from two neighbouring forests, one of which contained hardwood trees like beech and oak, and the other pine trees.

Month	CO_2 output $\mu l\ g^{-1}\ h^{-1}$	
	Hardwood leaf litter	Pine leaf litter
J	20	15
F	15	10
M	50	40
A	40	30
M	65	60
J	65	35
J	55	40
A	50	55
S	45	40
O	40	25
N	30	20
D	15	10

When these data are plotted as cyclical polygons, the result is as follows.

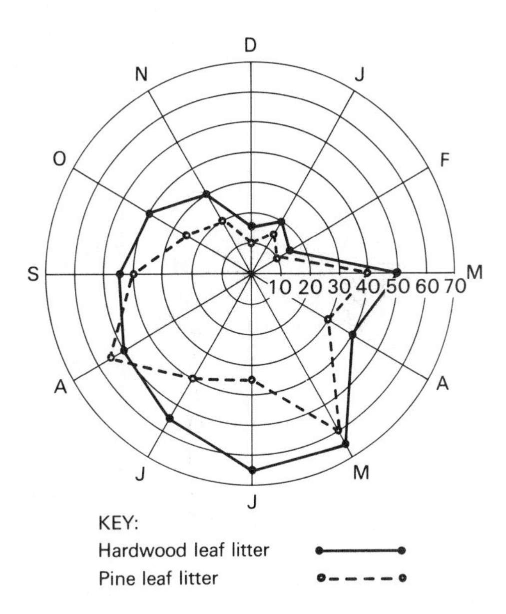

When they are plotted using bar graphs, the visual impact of the graph is again enhanced.

Both graphs would benefit from the use of colour, which would tend to separate the two sets of data more clearly.

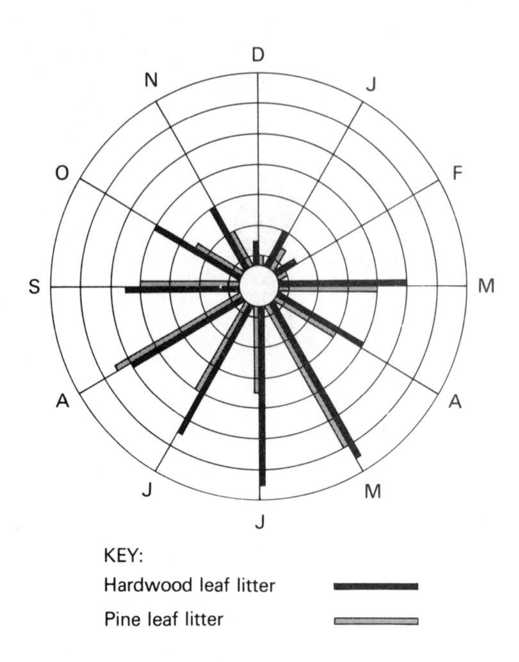

Example
Each month throughout the year the stomach contents of crows were examined and analysed. The results are shown.

Month	Percentage of food items in stomachs			
	Grain and seed	Fruit	Carrion	Invertebrates and small mammals
J	90	–	1	9
F	80	–	2	18
M	65	–	5	30
A	50	–	–	50
M	15	–	2	83
J	2	–	–	98
J	2	1	–	97
A	2	15	–	83
S	15	22	–	63
O	40	20	–	40
N	70	1	5	24
D	93	1	3	3

These data are displayed by means of a cyclical histogram showing the percentages of each type of food that the crows had eaten each month. The overall visual effect has been heightened by leaving out the grid.

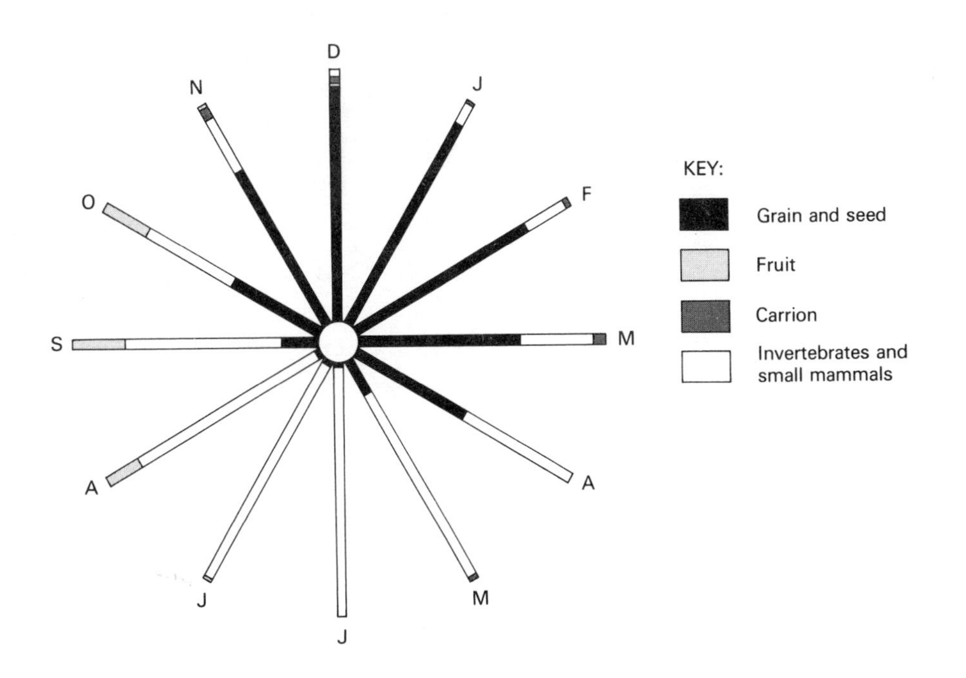

EXERCISE

The circular grid is also useful when presenting data which are related to direction, e.g. compass bearing. Ten trees were sampled to find out if the distribution of the single-celled alga *Pleurococcus*, which grows on the bark, was related to aspect. The following abundance scale was used in conjunction with a 10 cm × 10 cm transparent quadrat.

Abundance scale	0	1	2	3	4	5
Percentage cover	0	1–10	11–25	26–50	51–75	75–100

WORK SHEET 36

The results are given on Work Sheet 36.

1 Plot these results on the circular grid provided.

2 From which direction does the prevailing wind blow?

3 Which factors do you think determine the distribution of this alga on the bark of trees?

2.2.8 Triangular graphs

Triangular graphs are drawn on triangular co-ordinate grids which consist of equilateral triangles, similar to those found on isometric paper, each angle of the triangle being 60°. The lines can be spaced at whatever interval is required, but at every millimetre with the 5 mm and 10 mm lines heavier is the most common arrangement.

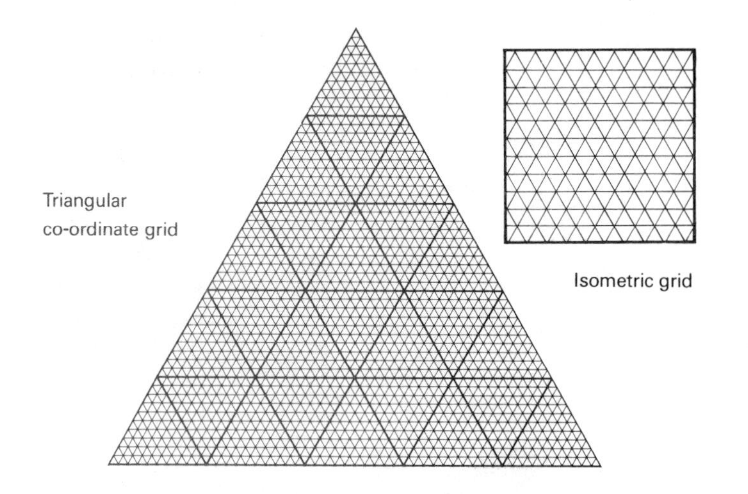

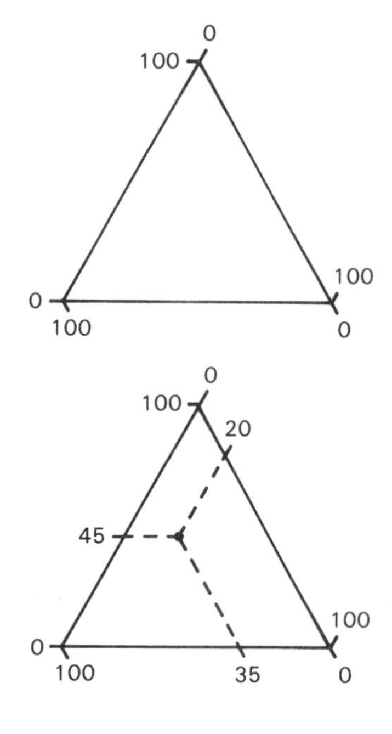

This type of graph enables three different variables to be considered; it has some similarities to three-dimensional line graphs and also to three-dimensional scattergraphs. The three variables must, however, be given as percentages, since this type of graph is used to show variations in proportions of a whole. Its main advantage is that it tends to highlight similar groups.

The three scales (0–100) run along each side of the equilateral triangle in a clockwise direction so that each of the points is 100 per cent for one scale and 0 per cent for the next scale.

Any combination of the three variables can be represented by a dot so long as they are given as percentages and the total adds up to 100, e.g. 20, 35 and 45 per cent.

Location of the intersection of only two of the variables is required since as all three variables add up to 100 (per cent), the third one will be predetermined (there are two degrees of freedom).

If the data are located in specific regions of the triangular graph then certain conclusions can be reached.

(i) If a dot is close to one of the points of the triangle, one of the components must be very large and two very small.

(ii) If the dot is located close to one side of the triangle, then one of components must be very small.

(iii) If the dot lies somewhere near the middle of the triangle, then all of the components are fairly evenly represented.

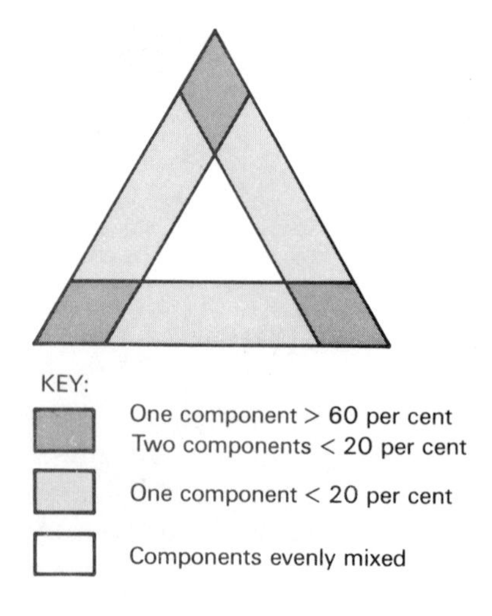

Example

Various foods were analysed to find the percentages they contained of carbohydrates, fats and proteins. The results are given in the table.

		Percentage		
		Carbohydrates	**Fats**	**Proteins**
Meat and Fish	**Liver**	5	23	72
	Beef	2	45	53
	White fish	21	26	53
Dairy products	**Milk**	42	33	25
	Ice-cream	55	34	11
	Yogurt	68	11	21
Nuts	**Peanuts**	9	58	33
	Almonds	5	68	27
	Coconut	8	83	8
Cereals and Veg.	**Wheat**	84	2	14
	Corn	86	3	11
	Potatoes	89	1	9

When the data are plotted as a triangular graph, the result is as below.

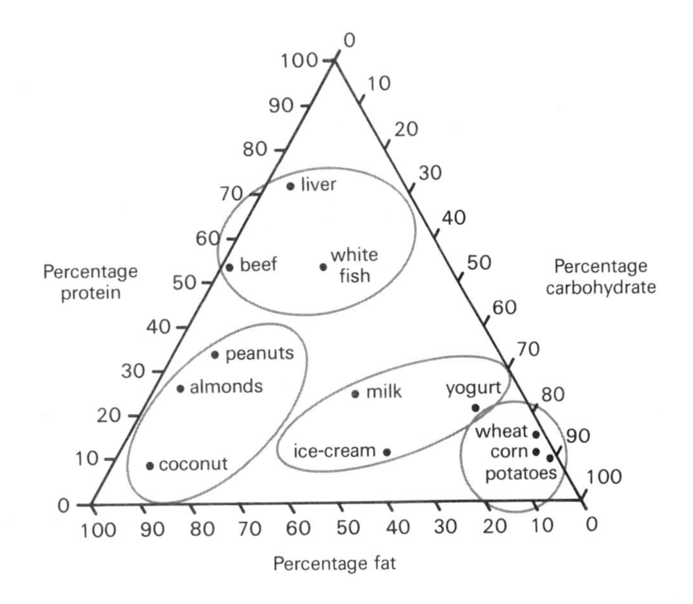

From the graph you can see that the points are located in four distinct groups. These tally with the four main food groups in the table:

(i) meat and fish;

(ii) dairy products;

(iii) nuts;

(iv) cereals and vegetables.

3 Analysing Data

3.1 GRAPHS AND PARAMETERS

3.1.1 Determining trends

Once graphs have been drawn of the data the picture becomes clearer – we can visualise the situation. The plotted points might suggest a trend – this is particularly important in Biology because the material is so variable. The graph will also give some idea about the relationship between the variables – this is equally important since we might want to find out how valid a hypothesis might be.

If the plotted points indicate a smooth curve then the interpolated lines should follow the curve. From such a graph intermediate values can be determined even though they might not be as exact as we might wish.

Drawing and analysing a compound line graph

EXERCISE
Try this exercise. You don't really need to know much about population genetics to do it.

The frequency of a gene D in a particular population is given as p, and the frequency of its recessive allele d as q.

1 Given that $p + q = 1$ and $p^2 + 2pq + q^2 = 1$, complete the table on Work Sheet 37 for values of p from 0 to 1 at intervals of 0.1.

WORK SHEET 37

You now have to construct line graphs using common axes on the linear grid provided (Work Sheet 37), to show the percentage frequencies of the three genotypes (DD, Dd, dd) plotted against the values of p *and* the values of q.

2 On which axis will you place the values of p (and q)?

3 Enter and label the values of p (from 0 to 1.0), and also the values of q (from 1.0 to 0), on the axis.

4 Write in the percentage points (0 to 100) on the other axis, and label it. What did you write as the label?

5 Plot the points for DD and draw a dashed line through them.

6 Plot the points for Dd and interpolate using a dot-dash line.

7 Plot the points for dd and interpolate using a continuous line.

From the graph determine the following:

8 The frequency of the genotype DD when $p = 0.78$.

9 The frequency of the genotype dd when $q = 0.65$.

10 The value of p when the frequencies of the genotypes DD and dd are equal.

11 The value of q when the frequencies of the genotypes Dd and dd are equal.

12 The value of p when the frequency of genotype Dd is at its maximum.

WORK SHEET 38

The data on Work Sheet 38 relate to changes in the respiratory quotient (RQ) of barley and flax over a 16-day period, which included germination and early growth of the seedlings.

1 Draw line graphs of both the barley and flax on the grid provided.

2 Label the lines where germination takes place.

3 Will the RQ for flax eventually equal that for barley? If so, on what day will this occur?

Using logarithmic grids to determine trends

In section 2.2.4 you learnt about the use of logarithmic scales when plotting graphs. Their main use occurs when the lines drawn on linear grids are exponential, i.e. the rate of a variable is demonstrated. The following are two further examples of their use.

Biston betularia

Speckled (peppered)

Melanic

EXERCISE

The introduction of industry with its associated smoke pollution darkened the natural environment. Speckled moths (*Biston betularia*), which rest during the day on the bark of trees, are well camouflaged. As the trees became darker over the years, natural selection favoured the darker or melanic forms.

The following data refer to each of ten successive years from the introduction of industry, when only 10 per cent of the moth population contained melanics.

Year	0	1	2	3	4	5	6	7	8	9	10
Percentage melanic forms	10	12	14.5	17.5	21	25	30	36	43	52	62

1 Plot these results on the linear grid on Work Sheet 39.

2 By extrapolation estimate the number of years it would take for the melanic forms to make up
(i) 90 per cent
(ii) 99 per cent of the population.

3 Now plot these results on the log × linear grid on Work Sheet 39.

4 Extrapolate the line and repeat the estimations as in question 2.

5 Which was the more accurate method?

WORK SHEET 40

Look back at the graph you drew concerning the frequencies of genes in a population (Work Sheet 37)

The values of p and the frequencies of DD (per cent) were as below.

p	0.1	0.2	0.3	0.4	0.5	0.6	0.7	0.8	0.9	1.0
Percentage DD	1	4	9	16	25	36	49	64	81	100

1 Plot these data on the two cycle × one cycle log grid on Work Sheet 40.

2 Is there any advantage in plotting on the log × log grid? Give reasons.

3.1.2 Determining relationships

Once graphs have been drawn and trends established, it is possible to express the relationship between variables in a quantitative manner. This is important when we want to confirm the validity of a hypothesis. Let us first look at line graphs on a linear grid.

Regression (type 1)

A problem often encountered in Biology is to find out if any relationship exists between variables, e.g. how does the body temperature of animals vary with changes in the environmental temperature?

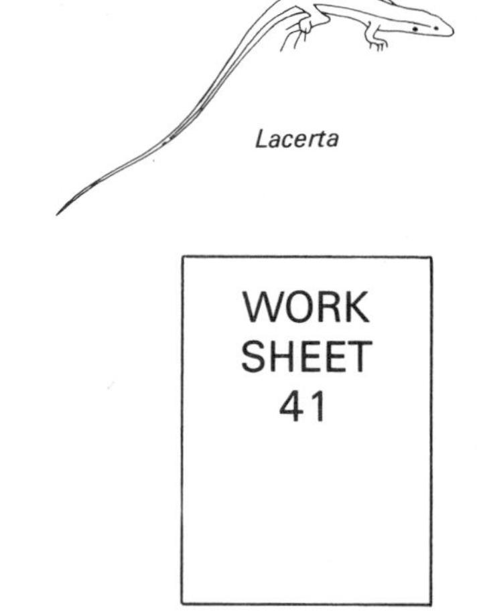

EXERCISE

To investigate this particular problem a reptile was subjected to a range of temperatures (5–40 °C). It was allowed to acclimatise at each temperature for 2 h, after which its body temperature was recorded. The results are given on Work Sheet 41. You can see that the body temperature matches the environmental temperature; this exact relationship is unlikely to occur in nature, it is given to illustrate the principle of regression. Because the relationship being investigated is that between an independent (X) variable and a dependent (Y) variable, we will call it regression type 1, to differentiate it from a second type of regression that you will be looking at later.

1 Which variable will occupy the X axis?

2 Starting each scale at a proper origin ($X0$, $Y0$) write in the scales for each axis and label.

3 Locate the points.

4 Using a ruler interpolate carefully.

The line slopes up as you move to the right, thus it has a ***positive gradient***, since the Y variable increases as the X variable increases. This line is known as the regression line.

Although we can tell from the table that $Y = X$, let us find out how this is derived from the graph. The equation for a straight line on ***linear grids*** is

$$Y = mX + C$$

where m is the gradient (steepness of the slope) and C is the value of Y when $X = 0$.

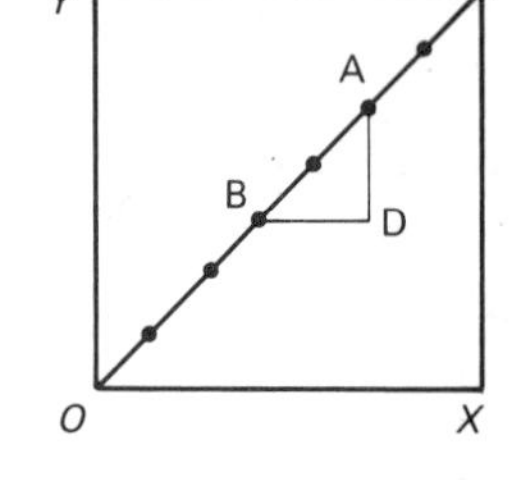

The value of m is determined by taking two points on the line and projecting along both the X and Y co-ordinates until they intersect. It doesn't really matter which two points you choose, but it is better to choose points which will make the calculation as straightforward as possible. On your plotted graph (Work Sheet 41) find point $X30, Y30$ and label it A. Draw a line down from point A to the X axis. Find point $X20, Y20$ and label it B. Draw a horizontal line from point B to intersect the line from point A. Label the intersection point D.

5 How many units of Y occupy the line AD?

6 How many units of X occupy the line BD?

To find the gradient of the line (the value of m) divide the number of units of Y by the number of units of X, i.e. AD/BD.

7 What is the value of m?

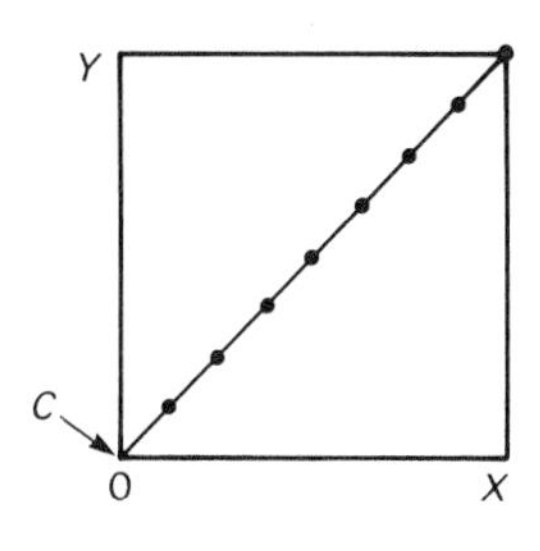

The value of C is determined by finding out where the regression line intersects the Yaxis, i.e. where $X = 0$. To find the value of C it is therefore necessary to have a proper origin, at least on the X axis. If displaced origins were used it would be impossible to determine the value of C from the graph.

8 What is the value of C?

9 Substitute your values of m and C in the equation $Y = mX + C$ to find the relationship between the X and Y variables.

WORK SHEET 42

In another experiment a particular species of protozoan was placed in varying concentrations of sea-water for one hour and the number of contractile vacuole expulsions was counted. The results are given on Work Sheet 42. From these measurements alone it would be difficult to work out an equation which relates these two variables. All we can say is that as the percentage sea-water increases the number of expulsions decreases.

1 Draw a graph of these results on the linear grid provided on Work Sheet 42. You should be able to choose the proper axes and arrange the scales to include an origin. Be careful when drawing the graph line to follow the instructions given in section 2.2.1.

2 Find the value of m. In this case the line descends to the right – it has a negative gradient so m will have a negative value.

3 Determine the value of C.

4 Substitute these values of m and C in the straight line equation to establish the relationship between X and Y.

5 Using your equation, calculate the value of Y when $X = 32$.

6 What is the value of Y when $X = 17.5$?

7 Check these calculated values on the graph. Are they confirmed?

Experiments were carried out on the effect of increasing the light intensity on the rate of photosynthesis of a particular plant. The table on Work Sheet 43 summarises the results, giving the rate of CO_2 exchange in mg CO_2 absorbed (+) or given off (−) per 50 cm^2 of leaf surface per hour. In this case since there is a minus value (−0.2) the X axis is moved up the grid so that a minus Y scale can be entered below the X axis. The position of the X axis is indicated on the grid to help you.

1 Plot a graph of the results on the grid provided on Work Sheet 43.

2 What sort of graph do you get?

3 From the graph develop an equation realting light intensity to CO_2 exchanged.

4 Using intermediate values of light intensity try out your equation, and from the graph find out if it is correct.

The log × linear graph and equation

In section 2.2.4 you looked at the log × linear graph. Examine the graph that you drew on this type of grid (Work Sheet 15). Since this graph turned out to be a straight line it indicated that there was some kind of relationship between the two variables – time and number of yeast cells.

WORK SHEET 44

EXERCISE

In a second experiment regarding the growth of yeast cells, a sample of the yeast culture was placed in a medium containing double the concentration of the nutrients that were present in the first experiment. A similar procedure was carried out and the results are given on Work Sheet 44.

1 Plot these results directly on to the log three cycle × linear grid provided. Label this line *Experiment 2*.

2 What sort of line did you obtain?

Since the line for Experiment 2 is steeper than the line for Experiment 1, the ***rate*** of growth of the yeast cells was greater in the second experiment. We would of course expect this since there was twice the concentration of nutrients. Both of the lines are straight, and you have already found that the relationship between the variables in such straight line graphs can be described by an equation.

The equation for a log × linear straight line is

$$Y = C \times m^X$$

Examine the line you have drawn from the results of Experiment 1 (Work Sheet 15). The value of C is where the graph line intersects the Y axis, i.e. where $X = 0$.

3 What is the value of C?

As in the linear grid, m represents the ***gradient*** of the line. In the case of the log × linear grid, however, the graph is a ratio graph, so we have to determine the ***ratio*** of Y to X. To do this divide any value of Y with its value one unit of time earlier, e.g. the value of Y when X is 1, divided by the value of Y when X is 0, i.e. 60 divided by 50.

$\frac{60}{50} = ?$, similarly $\frac{72}{60} = ?$, and $\frac{86.4}{72} = ?$, etc.

4 What therefore is the value of m?

5 Substitute the values of C and m in the equation and find the relationship between the X and Y variables.

6 Check that the equation is correct by calculating the value of Y when $X = 2$ h.

Once you have established that the equation is correct, you can use it to calculate the number of yeast cells after any given number of hours.

7 What would be the number after 7 h?

Now examine the line for Experiment 2.

8 What is the value of C?

9 What is the value of m?

10 What is the relationship between X and Y?

11 Calculate the number of yeast cells after 7 h.

Compare the values of Y that you have calculated in questions 7 and 11 with those estimated from the graphs.

12 Do the values agree?

You should have found that the value of m in Experiment 2 is twice that of m in Experiment 1. This indicates that the rate of growth of the yeast in

Experiment 2 was double that of the yeast in Experiment 1. We might expect something like this since there was double the concentration of nutrients.

We have extrapolated the lines without any real justification. Would the population continue to grow at the same rate for 14 h? This would depend on a number of factors.

13 Suggest what these factors might be.

Go back to section 3.1.1 (p. 51), where you drew a log/linear graph of years against the percentage of melanic moths in a population (Work Sheet 39). Since the result was a straight line, the relationship between the two variables should be quantifiable.

In this case you are being asked to determine X from Y, not Y from X. Since the log/linear equation is $Y = C \times m^X$, then $X = \frac{\log Y - \log C}{\log m}$ (log values can easily be determined using a calculator).

14 Use this equation to determine the number of years after which
(i) 90 per cent
(ii) 99 per cent
of the population are melanic.

15 Do your answers to question 14 agree with those you determined from the graph?

The log × log graph and equation

WORK SHEET 45

EXERCISE

In an experiment to determine the effect of changes in light intensity on the rate of oxygen production in *Elodea* using a modified 'Audus' apparatus, it was decided to use the distance of a lamp from the *Elodea* as a measure of the light intensity. The lamp was placed at varying distances from a photoelectric cell and meter readings were taken. The results are given on Work Sheet 45.

1 Draw a graph of these results on the log two cycles × log one cycle grid provided on Work Sheet 45.

Since this is a straight line there must be a relationship between the light intensity and the distance of the lamp from the plant; the line descends to the right, indicating that as the distance increases, the intensity decreases, obvious enough. The gradient m is therefore negative. The gradient is estimated by measuring actual distances along the X and Y axes at any particular plotted point, e.g. $X = 10$, $Y = 1000$, which in this case will be the length of the X axis and the length of the Y axis. The gradient m will therefore be

$$\frac{\text{the length of the } Y \text{ axis}}{\text{the length of the } X \text{ axis}}$$

2 What is the value of m?

In the case of the log × log graph, the value of C can be determined by using the equation $\log Y = m \times \log X + \log C$. By substituting known values of X and Y in this equation, we can determine the value of C since we have already found the value of m.

3 Substitute the values of $X = 10$ and $Y = 1000$ (since when $X = 10$, $Y = 1000$) in the equation and determine the value of C.

The equation for ***log × log*** graphs is $Y = C \times X^m$.

4 Substitute your values of m and C in the equation to determine the relationship between lamp distance and light intensity.

5 Check your equation by letting $X = 50$. Is the answer correct?

6 Using the equation, calculate the light intensities for lamp distances of 20, 40, 60 and 80 cm.

7 Check these values on the graph. Are they correct?

Correlation

In section 2.2.5 you saw that the correlation of two interdependent variables described the intensity of the relationship between them. In what way is the correlation quantified?

Once the scattergraph has been constructed an examination of the distribution of the dots can indicate whether there is any relationship between the two variables – the closer the dots came to lying on a straight line the closer the relationship, the more scattered the dots the less close the relationship. If the dots exhibit a trend, we are justified in thinking that there is some sort of relationship. If the points lie on a perfectly straight line then the correlation is perfect; if the line slopes up from left to right the correlation is positive; if the line drops from left to right the correlation is negative.

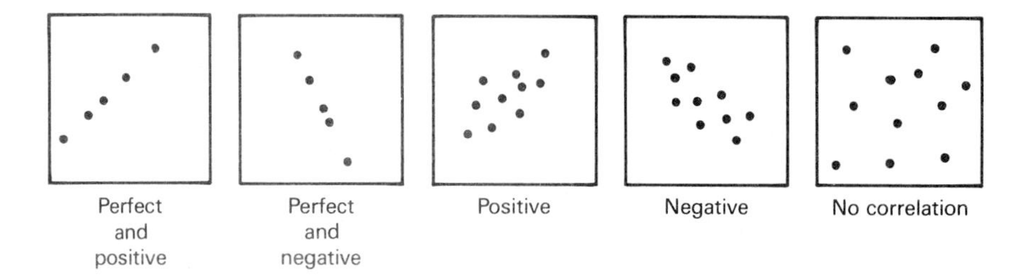

Since the material used in Biology is so variable the dots are usually scattered and only trends may be discerned.

The correlation coefficient is symbolised 'r'. The values of r range from $+1$ to -1; $+1$ means a perfect positive correlation; 0 means no correlation, and -1 means a perfect negative correlation. Intermediate values of r correspond to distributions whose scattergraphs are intermediate.

EXERCISE

On Work Sheet 17 you plotted a set of hypothetical data. Examine this scattergraph.

1 Is the correlation positive or negative?

The pattern of dots gives us some idea of the degree of correlation existing between the two variables. We might even be able to quantify the correlation approximately, i.e. give r a value. The following diagrams exhibit a range of scattergraphs with their associated r values. Examine them carefully to find the connection between the r value and the appearance of the scattergraph.

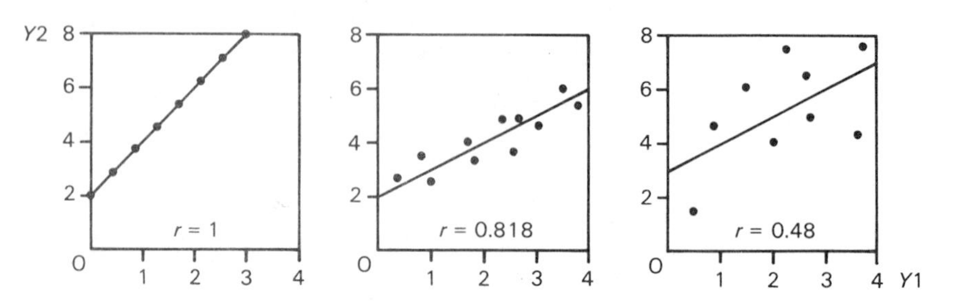

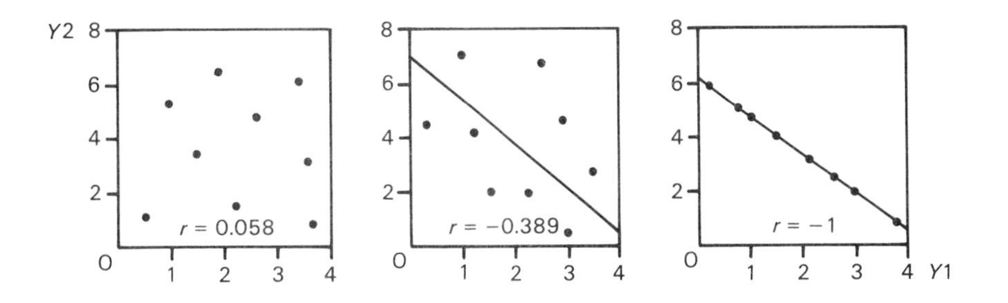

2 Estimate the value of r from the scattergraph on Work Sheet 17.

Regression (type 2)

You saw on p. 52 that the relationship between the X and Y variables could be determined using the regression line and straight line equation. In the case of the scattergraph, if the pattern of dots indicates a possible relationship between the variables (positive or negative) then a straight line can be drawn through the dots which most nearly fits all the points. This is the ***regression line*** or the ***line of best fit***. Since in this case we are dealing with the interdependence of two variables we call this regression type 2 to differentiate it from the independent/dependent situation which is type 1.

Although we used $Y1$ and $Y2$ for the two interdependent variables when drawing the scattergraph, it becomes confusing when we start to use formulae so the only thing to do is to revert to the X/Y situation.

WORK SHEET 46

EXERCISE

1 Using the data given on Work Sheet 46, draw a scattergraph on the grid provided.

How is the line of best fit determined? There are a number of possible techniques that can be employed.

Method 1: Place a transparent ruler on the scattergraph and move it around until you think that the edge passes through the middle of the dots. Then draw a straight line through them.

Method 2: On the same scattergraph draw ***two*** parallel lines which enclose most, if not all, of the dots. Then draw a straight line equidistant between the two parallel straight lines.

2 Does the line drawn using method 2 coincide with the line you drew using method 1?

Method 3: Calculate the mean (average) of the X values ($\overline{X}$) and the mean of the Y values ($\overline{Y}$). Using these mean values as coordinates plot the point $\overline{X}, \overline{Y}$ with a distinct cross (×). For the definition of the mean see p. 67 and Appendix 1.

3 Does this point × lie on the line that you have drawn using method 2? Join the point × to the origin ($X = 0,\ Y = 0$) by a straight line. This line is a good approximation to the line of best fit (regression line).

It is important that the regression line should relate as closely as possible to both m and C.

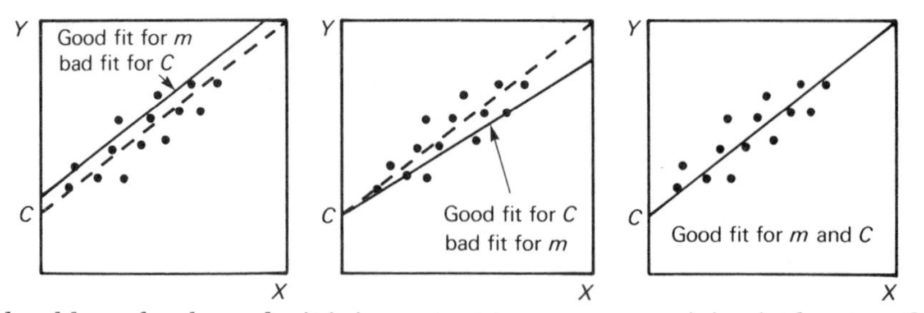

It should now be clear why it is important to use proper origins (at least on the X axis), since otherwise it would be impossible to obtain a value for C. To find the gradient m of the regression line, the usual procedure is carried out.

Two points, A and B, are selected towards each end of the regression line. These points should be selected so that they are as convenient as possible for the purposes of calculation, but keep them well apart. Once the regression line has been established the straight line equation can be used. From p. 52 you will (or should!) remember that this equation is $Y = mX + C$, where m is the gradient and C the intercept with the Y axis (the value of Y when $X = 0$).

Method 4: A reasonably accurate method which is appropriate when displaced origins are used was devised by Bartlett, an English statistician. To find the 'best fit' straight line:

(i) and (ii)

X	Y	
5.1	42	Group 1
5.3	48	
5.5	40	
5.7	47	
6.0	40	
6.1	52	Group 2
6.2	47	
6.4	55	
6.8	48	
6.9	56	
7.2	62	Group 3
7.3	53	
7.5	61	
7.6	53	
7.9	58	

(i) The observations are arranged in ascending order with respect to one of the variables, e.g. the X variable.

(ii) These ranked observations are then divided into three groups of equal size. If the sample can't be divided into three equal groups, then at least the lowest and highest groups should be of equal size.

(iii) The means of X and Y for the whole sample are computed and represented by $\overline{X}$ and $\overline{Y}$.

(iv) The means for the lowest (1) and the highest (3) groups are then computed. These can be represented by $\overline{X}_1$ and $\overline{Y}_1$ for the lowest groups, $\overline{X}_3$ and $\overline{Y}_3$ for the highest groups.

(v) The points $\overline{X},\overline{Y}$; $\overline{X}_1,\overline{Y}_1$; $\overline{X}_3,\overline{Y}_3$ are plotted as crosses on the scattergraph and the three points are joined by a straight line. This is the line of best fit.

Let us find out how Bartlett's method works with the example given.

In this example there are fifteen observations, so there are three groups of five.

(iii) Mean of X, i.e. $\overline{X} = 6.5$
Mean of Y, i.e. $\overline{Y} = 50.8$

(iv) Mean of X in group 1, i.e. $\overline{X}_1 = 5.52$
Mean of Y in group 1, i.e. $\overline{Y}_1 = 43.4$

Mean of X in group 3, i.e. $\overline{X}_3 = 7.5$
Mean of Y in group 3, i.e. $\overline{Y}_3 = 57.4$

To determine the gradient m:

$$m = \frac{\overline{Y}_3 - \overline{Y}_1}{\overline{X}_3 - \overline{X}_1} = \frac{57.4 - 43.4}{7.5 - 5.52} = \frac{14}{1.98} = 7.07$$

Since there are no proper origins the intercept C is determined using the expression $C = \overline{Y} - m\overline{X}$

$$\text{so } C = 50.8 - (7.07 \times 6.5) = 4.85$$

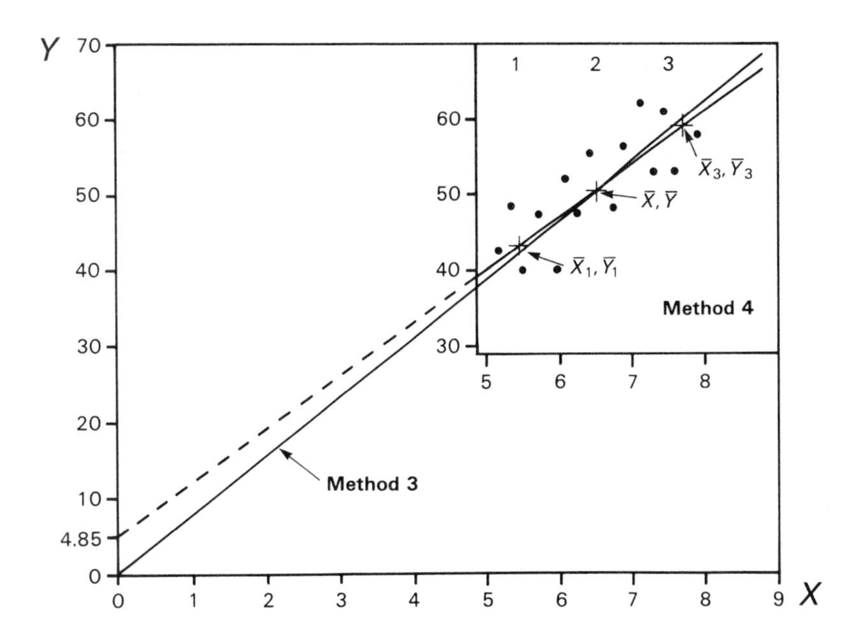

The proper origins are shown on the graph to illustrate how the projected line crosses the Y axis at a value of 4.85, as well as illustrating Bartlett's method.

When methods 3 and 4 are compared on the graph, we see that the two lines are quite close together but method 4 is the most accurate. This is therefore the one you would be advised to use.

The relationship between the two variables X and Y has therefore been established: $Y = mX + C$

Since $m = 7.07$ and $C = 4.85$;

then $Y = 7.07X + 4.85$

WORK SHEET 47

EXERCISE

The data on Work Sheet 47 concerns the length and breadth of 25 privet leaves.

1 Draw a scattergraph of the data on the grid provided.

2 Draw in the line of best fit (the regression line) using the most accurate method; but try all four for practice.

3 Determine m and C.

4 Determine the relationship between length and breadth in privet leaves.

The correlation coefficient *r*

In order to determine whether a particular correlation is significant or not it is necessary to have a more exact value. The correlation coefficient, designated 'r', is determined using the following formula:

$$r = \frac{\Sigma[(X - \overline{X})(Y - \overline{Y})]}{\sqrt{[\Sigma(X - \overline{X})^2 \Sigma(Y - \overline{Y})^2]}}$$

where X is the first variable;
$\overline{X}$ is the mean (average) of the X variates;
Y is the second variable
$\overline{Y}$ is the mean of the Y variates.

This is quite an involved formula to compute, particularly when a large number of pairs of data are involved; in such cases it is useful to use a programmable calculator or a computer.

Example

Using a small sample let us find out how to calculate r.

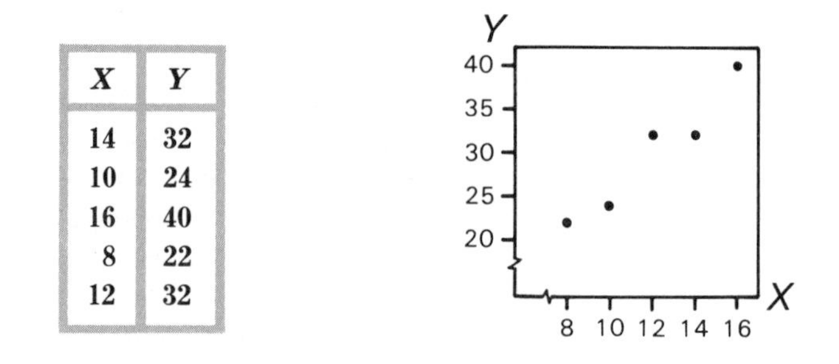

X	Y
14	32
10	24
16	40
8	22
12	32

When plotted as a scattergraph we can see that the correlation is positive and r is quite high, being probably in the region of 0.9. Let us find out the actual value of r.

Always try to arrange the work in an organised way, using suitable columns.

X	Y	$(X-\bar{X})$	$(Y-\bar{Y})$	$(X-\bar{X})(Y-\bar{Y})$	$(X-\bar{X})^2$	$(Y-\bar{Y})^2$
14	32	2	2	4	4	4
10	24	−2	−6	12	4	36
16	40	4	10	40	16	100
8	22	−4	−8	32	16	64
12	32	0	2	0	0	4
$\sum = 60$	$\sum = 150$			$\sum = 88$	$\sum = 40$	$\sum = 208$

$\bar{X} = 60/5 = 12$ $\quad \bar{Y} = 150/5 = 30$

Substituting in the formula, $r = \dfrac{88}{\sqrt{40 \times 208}} = \dfrac{88}{91.2} = 0.96$

This value confirms our estimate that the correlation was high and we can see from the scattergraph that it is positive. ***Always draw the scattergraph and roughly estimate r before calculating*** it so that you have a rough check on your calculation. To find out if this value of r is significant, go to section 3.3.2 p. 111.

EXERCISE

1 Compute the value of r for the data given on Work Sheet 17 (read X for $Y1$ and Y for $Y2$).

2 Does this value agree with your approximation given in answer to question 2 on p. 56?

3 Compute r for the data on Work Sheet 46.

3.1.3 Frequency distributions

There are a number of different types of frequency distributions, all of which possess certain important characteristics. When we draw graphs of frequency distributions the patterns which emerge should give us clues about the distribution. Often the patterns can suggest mathematical relationships in the same way that the line graphs did.

The normal distribution

Many, if not most, of the characteristics with which a biologist deals follow Quetelet's Principle, which states that 'certain classes of the variable occur more frequently than others, and the frequency become progressively less as the classes are farther in either direction from these most common values'. Examples of such characteristics are the heights of people, the weights of rats, the number of flowers on plants, the number of peas in a pod, the body temperature of mice, the breathing rate of locusts, the oxygen consumption of humans. The list is endless – in fact almost all anatomical and physiological characteristics approximate to Quetelet's Principle. Because of the universal nature of this type of distribution (the 'normal' distribution) a mathematical model in the form of an equation has been worked out which relates the frequencies of the different values of a variable to their size.

When the normal distribution equation is drawn as a graph it assumes a very characteristic 'bell' shape. Since this is a theoretical distribution based on an infinite number of observations, it is unlikely that the distribution curve plotted for most data will conform exactly to this curve – it will generally only approximate to it.

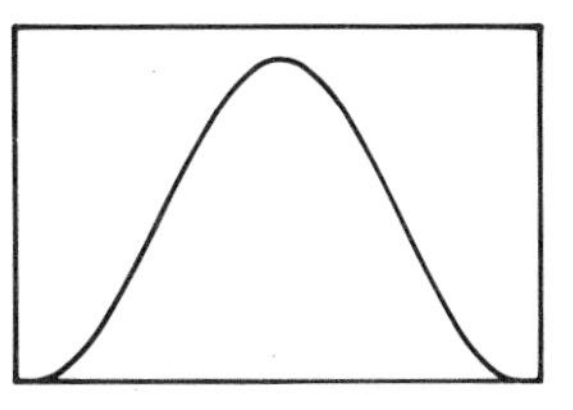

Testing the normality of a distribution

We saw that in the case of line graphs it is useful to convert curved graph lines into straight lines since this makes interpretation not only easier but also more accurate.

In a hypothetical sample of 435 plants, the number of flowers on each plant was recorded and the results displayed as a frequency table.

No. of flowers per plant	f
8–12 (10)	1
13–17 (15)	10
18–22 (20)	40
23–27 (25) (mid-points	100
28–32 (30) of classes)	135
33–37 (35)	100
38–42 (40)	40
43–47 (45)	10
48–52 (50)	1

From these data the cumulative frequencies are worked out and displayed.

When these data are plotted as a histogram the result is as follows: (see next page).

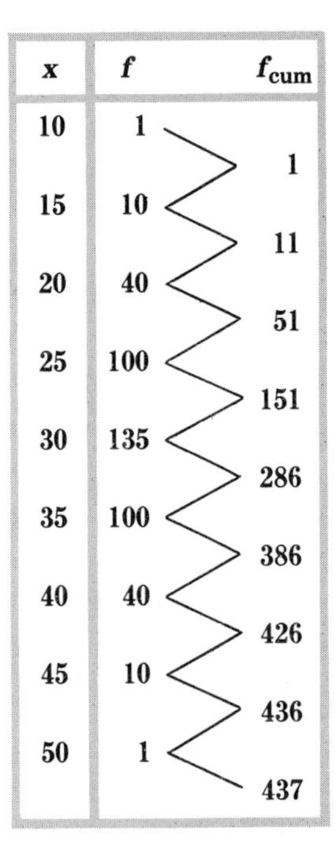

x	f	f_{cum}
10	1	1
15	10	11
20	40	51
25	100	151
30	135	286
35	100	386
40	40	426
45	10	436
50	1	437

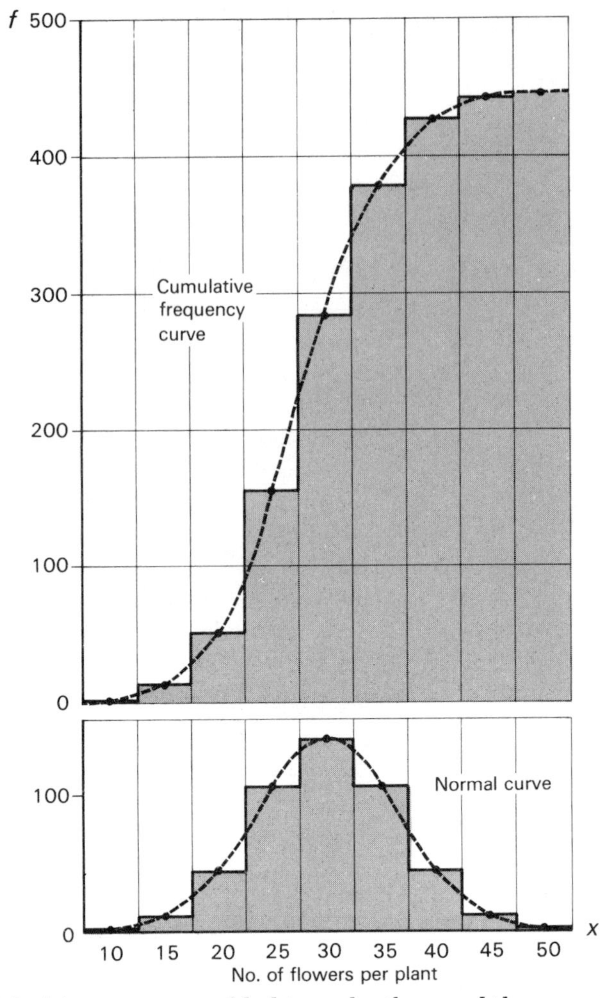

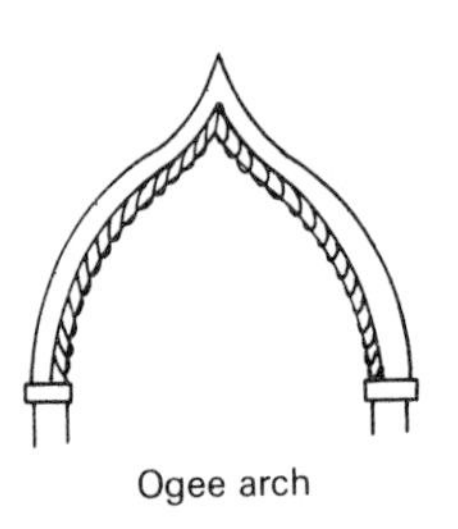
Ogee arch

The bars of the histogram are added to each other, and the curve is formed by joining up the mid-points of the classes. The curve assumes a characteristic 's' or sigmoid shape and is known as a ***cumulative frequency curve***. Because of their shape such curves are sometimes known as 'ogive' curves, since they are similar to one-half of an Ogee arch (Gothic), the Old French term being Ogive.

Usually the cumulative frequency curve has the *f* axis represented not by actual numbers, but by percentages. The cumulative values must therefore be translated into percentages.

x	f	f_{cum}	Percentage f_{cum}
10	1	1	0.23
15	10	11	2.52
20	40	51	11.70
25	100	151	34.60
30	135	286	65.40
35	100	386	88.30
40	40	426	97.50
45	10	436	99.80
50	1	437	100.00

Such an arrangement of the data can be used to determine how closely the distribution resembles the 'normal' distribution: the closer it is, the nearer it will approximate to a straight line when plotted on a probability grid.

Probability grids

In probability grids, one or both of the axes consists of a probability scale, which corresponds to a cumulative normal distribution curve. For particular uses various combinations of axes are possible, e.g. probability × linear, probability × probability and probability × logarithmic (any number of cycles).

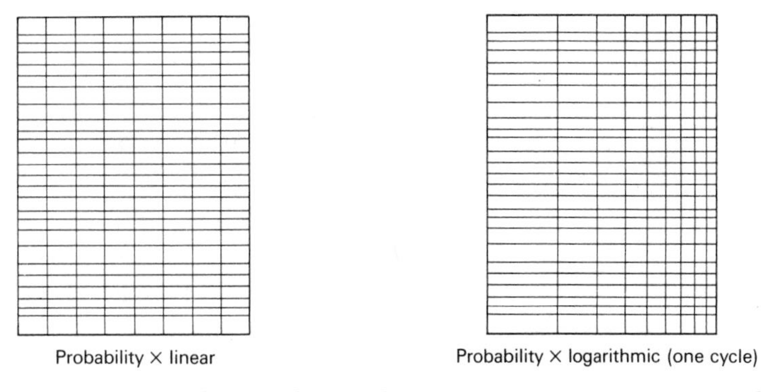

Probability × linear   Probability × logarithmic (one cycle)

The data concerning the number of flowers on a certain species of plant are plotted on a probability grid as below. As you can see, the interpolated line is straight. Using such an arrangement not only makes interpolation more straightforward and accurate, but also tells us that this particular distribution is a normal distribution.

As you will see later (section 3.1.4), this use of probability grids can be very helpful in the estimation of various statistical parameters.

x	f	f_{cum}	Percentage f_{cum}
10	1	1	0.23
15	10	11	2.52
20	40	51	11.70
25	100	151	34.60
30	135	286	65.60
35	100	386	88.40
40	40	426	97.50
45	10	436	99.80
50	1	437	100.00

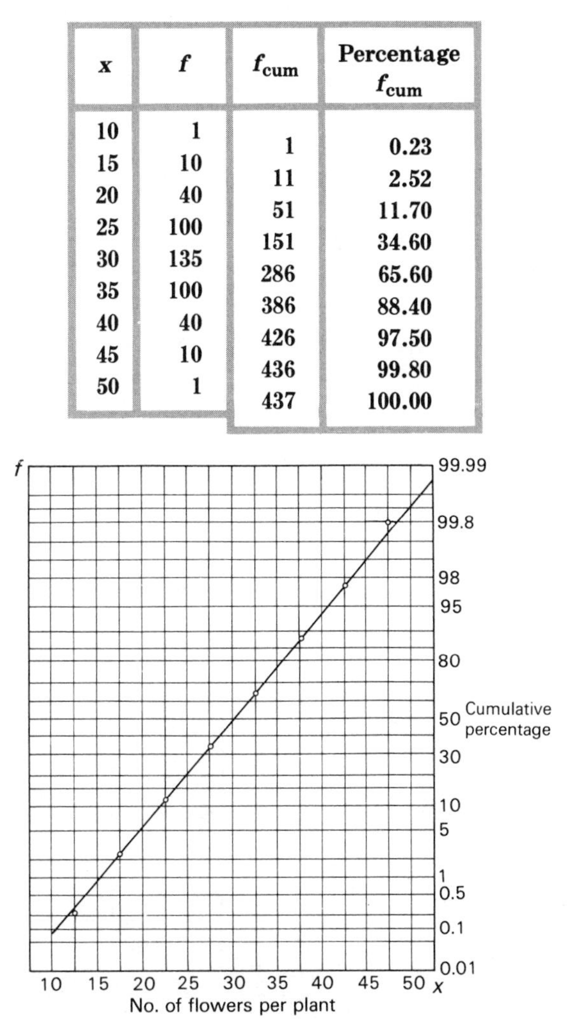

The binomial distribution

Often in Biology we merely classify individuals into two categories which are quite separate and distinct, e.g. normal or vestigial wing, seeds germinated or dormant, something or other present or absent. The frequencies with which individuals fall into the two categories are then determined (and may be given as a proportion or percentage). Data of this sort tends to follow the binomial expansion and the frequency distribution of such data is thus known as a binomial distribution.

Example

In genetics the characteristics are often of an either/or type, quite clearly distinguished from one another.

The ratio obtained in the F2 generation from a monohybrid cross is 3:1, for a dihybrid cross 9:3:3:1, for a trihybrid cross 27:9:9:9:3:3:3:1, etc. Such ratios form a binomial expansion. Since such expansions are described by Pascal's triangle, let us see how this is related to the F2 ratios. We can see that the ratio figures are multiples of 3, so that each value can be described in terms of 3, i.e. $1 = 3^0$, $3 = 3^1$, $9 = 3^2$, $27 = 3^3$, etc. When these are arranged in tabular form the result is as follows.

Number of characters	3^8	3^7	3^6	3^5	3^4	3^3	3^2	3^1	3^0	$3^0 = 1$; $3^1 = 3$; $3^2 = 9$; $3^3 = 27$ etc.	Ratio
1								1	1	1 of 3;1 of 1	3:1
2							1	2	1	1 of 9;2 of 3;1 of 1	9:3:3:1
3						1	3	3	1	1 of 27;3 of 9;3 of 3;1 of 1	27:9:9:9:3:3:3:1
4					1	4	6	4	1	etc.	etc.
5				1	5	10	10	5	1		
6			1	6	15	20	15	6	1		
7		1	7	21	35	35	21	7	1		
8	1	8	28	56	70	56	28	8	1		

The emerging pattern is Pascal's triangle (slightly modified). When such distributions are plotted as histograms they look like those below.

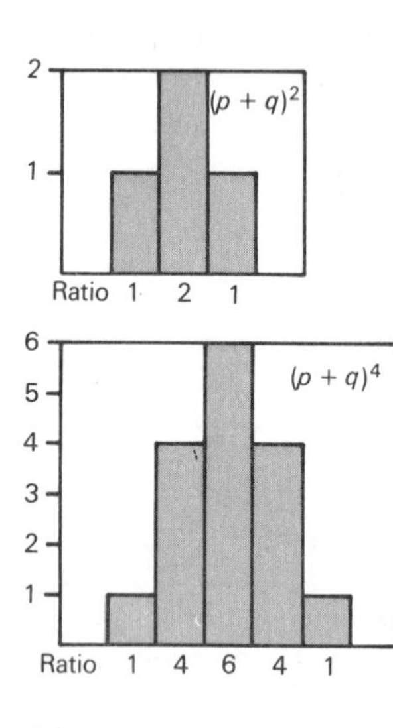

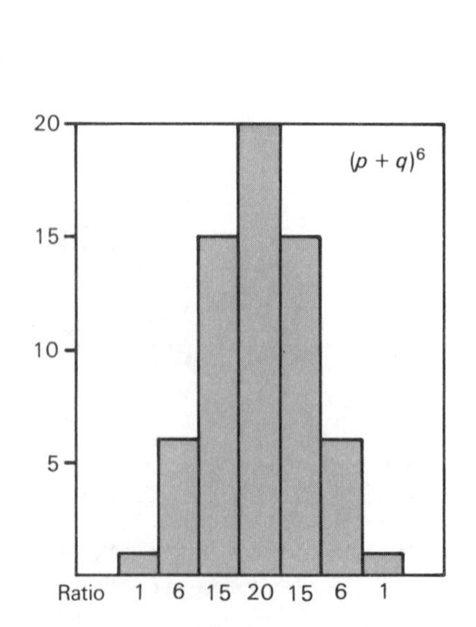

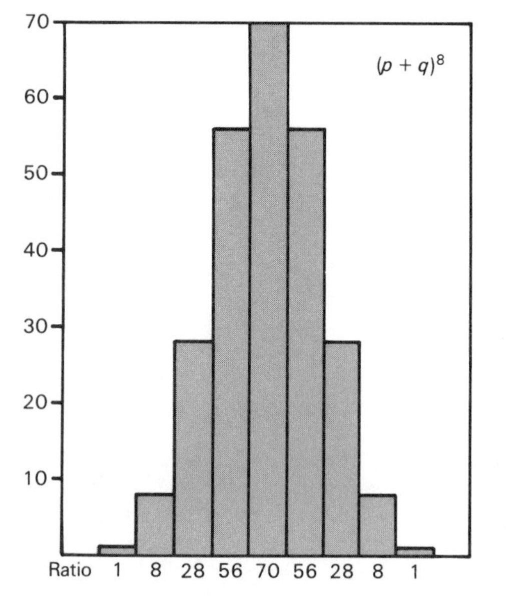

You can see that as the numbers of characters increase, the more closely the distribution approaches the 'normal' distribution. In fact, such distributions can, for all intents and purposes, be dealt with in essentially the same way as the normal distribution. Similar distributions also occur in characteristics that are controlled by a number of genes (polygenic characteristics).

The Poisson distribution

This type of distribution is named after the French mathematician Poisson (1781–1840). When we count randomly distributed items or events of an unbiased sample they will tend to form a Poisson distribution. This type of distribution is in fact an approximation to the binomial distribution, but it is asymmetrical. Since the Poisson distribution occurs in situations where events occur in a random way it can be used to provide evidence as to whether or not true randomness is present.

3.1.4 Statistical parameters

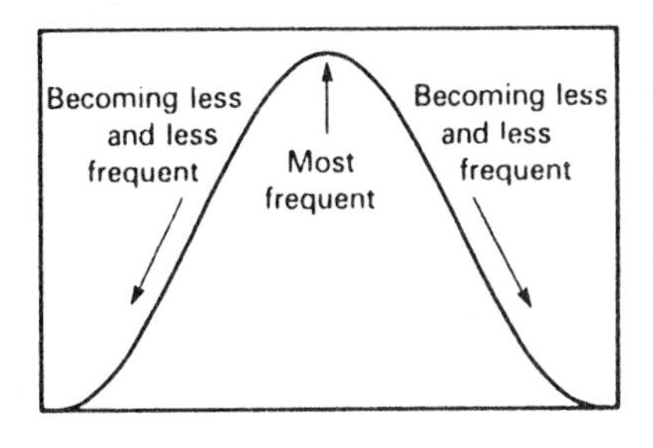

When dealing with data that are concerned with populations and samples from populations, certain parameters are of particular value. The normal distribution, which is based on a theoretical model, possesses particular values for these parameters so that we can compare sample or population parameters with them. You have already seen that in distributions conforming to the normal distribution, the variates are more frequent around some particular value and they become less frequent as you move from this value in either direction.

When examining any distribution we need therefore to quantify

(i) the point around which the variates tend to cluster, and

(ii) the extent to which the variates spread out from this point.

Measures of central tendency

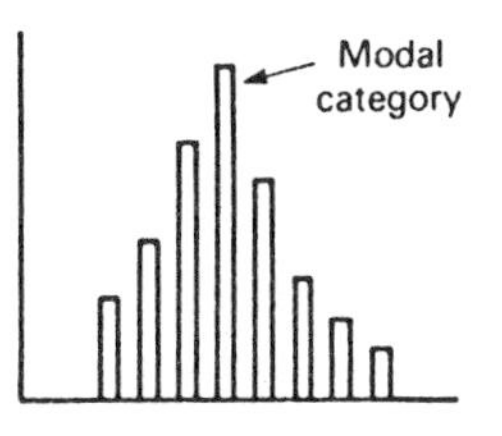

There are three kinds of measure of the 'central number' of any distribution.

The mode

The mode of a frequency distribution is the most common class. When a histogram is drawn, it is the class with the largest area, but since the columns are usually drawn of equal width, it will normally be the class with the tallest column (height is proportional to the area).

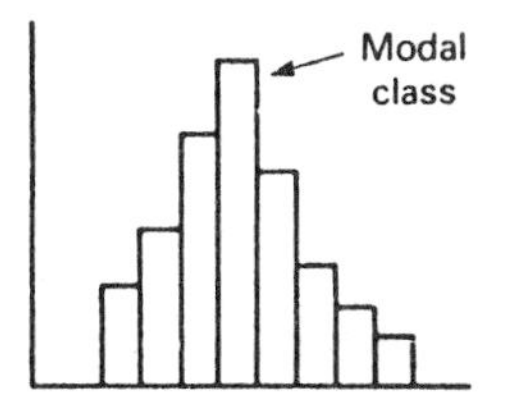

The mode is the simplest measure of the central number. It is the 'most common' category or value of a variable – it occurs most often in a frequency distribution. The mode is relatively easy to find, providing the distribution is ***mono-modal***, i.e. there is only one modal value. In such cases the bar-graph or histogram will have a peak which corresponds to the modal category or class.

EXERCISE

The data on Work Sheet 48 refer to the length of limpets (*Patella vulgata*) collected at mean tide level on a rocky shore.

1 Draw a histogram of the data on the linear grid provided (Work Sheet 48).

2 From the graph, which class is the most common, i.e. the modal class?

You will notice that the classes immediately above and below the modal class are of unequal frequency, one being 12 and the other 22. The actual modal

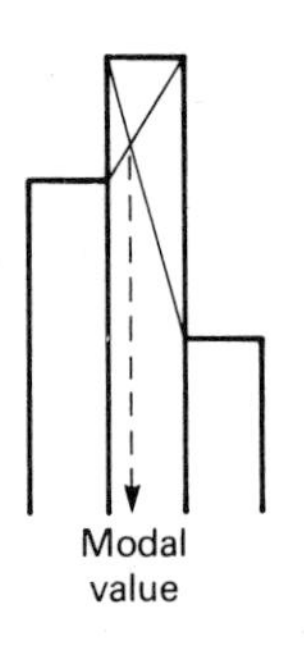

value is not therefore the mid-point of the modal class, but lies to the left of this value because of the effect of the class just to the left of the modal class.

To overcome this problem join up the points shown in the diagram and project down to the X axis from the intersection of the lines. This will be very close to the modal value.

3 What is the modal value of the data?

In section 2.2.6 you drew a number of bar-graphs and histograms. Look back again at them and find the modal categories and classes. The mode is not used very often in Biology, but it is the only measure of the average that can be used for a variable consisting of categories.

The median

This measure of central tendency is particulary useful when dealing with categories or values that can be ranked. It is defined as the value which, if the observations are placed in rank order, will divide the distribution into two halves.

Example

Data set	**8 4 6 10 2 12**
Ranked	**2 4 6 8 10 12**

In this example the median will lie between the third and fourth observations, i.e. between 6 and 8 – conventionally it is taken to be halfway between these, i.e. 7. If there is an odd number of observations, e.g. 2, 4, 6, 8, 10, then the median will be the value in the middle, i.e. 6.

When frequency distributions are involved the median can be found by using the cumulative frequency curve (p. 62).

Example

Look again at the curve obtained from the flower data (p. 62).

x	f	f_{cum}
10		
15	1	1
20	10	11
25	40	51
30	100	151
35	135	286
40	100	386
45	40	426
50	10	436
	1	437

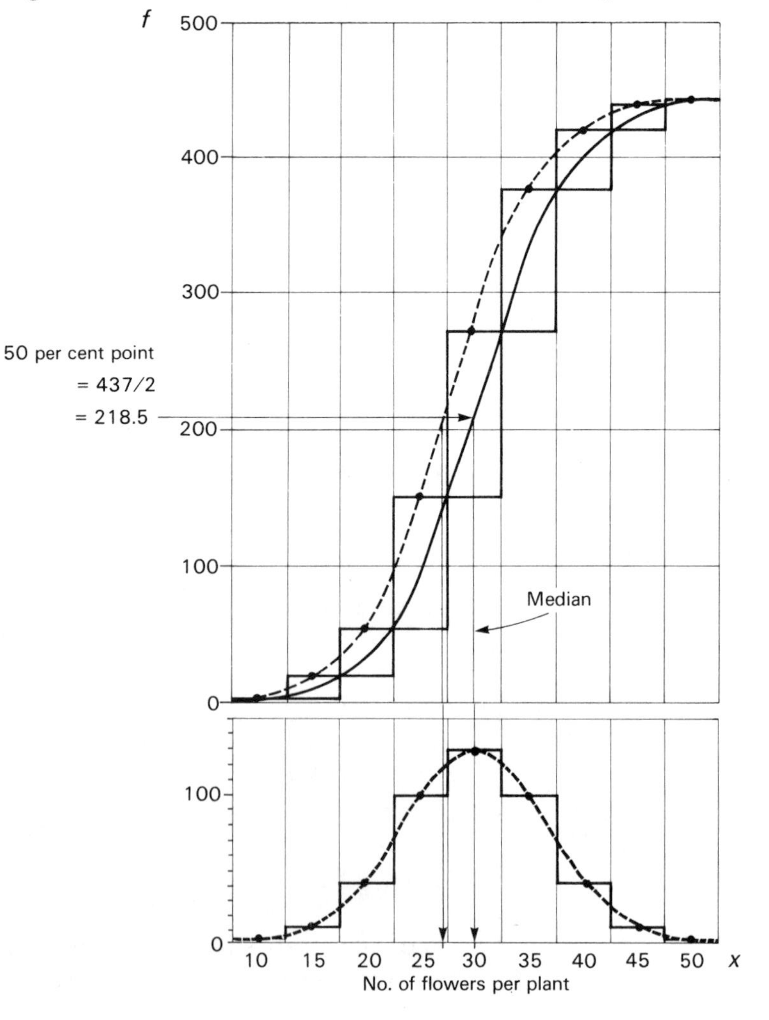

The total sample consists of 437 readings, so the reading 437/2 will have the median value, 437/2 = 218.5. If you read across from the 218.5 point until you intersect with the dotted line and then project down to the X axis you will find that the normal distribution at the bottom is not divided equally in half.

In this example, since the distribution is symmetrical, the median should divide the distribution exactly in half, but it doesn't! Why is this?

The mid-points of the columns in the cumulative frequency curve are not a true representation of the gain at each step. If you look at the table you will see that the cumulative values lie between the mid-points. A truer picture of the gain can be achieved by joining up the diagonals of the steps, as shown by the solid line (the points are located half an interval to the right). The projection from 218.5 now intersects the solid line at a value of 30, which is the median.

We saw on p. 63 that the data produced a straight line when plotted on a probability grid, using the percentages of the cumulative frequencies. From such a graph we can easily determine the median – it will be the value of x when f_{cum} = 50 per cent.

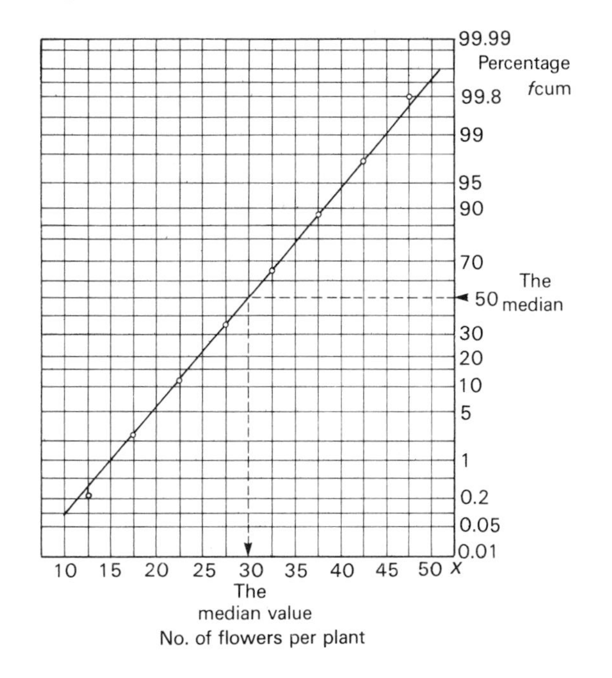

The mean

This is the most widely used measure of central tendency. It is properly known as the 'arithmetic mean' but is usually shortened to just the 'mean'. The mean is what we often refer to as the average. It is determined by adding together all the observed values and dividing by the number of observations.

The formula is $\bar{x} = \frac{\Sigma x}{n}$ where $\bar{x}$ (pronounced 'x bar') is the mean; Σ (sigma) is 'the sum of'; x includes all the variates, and n is the number of variates.

Example

x = 9, 11, 18, 44, 106, 938, 124, 67, 22 and 12
Σx = 9 + 11 + 18 + 44 + 106 + 938 + 124 + 67 + 22 + 12 = 1351
n = 10, so $\bar{x}$ = 1351/10 = 135

Such a value can easily be computed using a calculator, many of which have an $\bar{x}$ function.

For grouped discrete data the mean is computed using the formula

$\bar{x} = \frac{\Sigma fx}{\Sigma f}$, where f is again the frequency.

Example

x	f	fx
2	4	8
3	5	15
4	10	40
5	29	145
6	30	180
7	18	126
8	3	24
9	1	9
	$\Sigma f = 100$	$\Sigma fx = 547$

$$\bar{x} = \frac{547}{100} = 5.47$$

If the data are grouped into classes because the variates are continuous, the mid-point value of each class is used.

Example

x	Mid-points of x	f	fx
35–40	37.5	7	262.5
40–45	42.5	18	765.0
45–50	47.5	23	1092.5
50–55	52.5	24	1260.0
55–60	57.5	16	920.0
60–65	62.5	12	750.0
		$\Sigma f = 100$	$\Sigma fx = 5050.0$

$$\bar{x} = \frac{5050}{100} = 50.5$$

Often the calculation of the mean can be simplified by using what is known as a false origin. If, for example, we wanted to find the mean of the values 997, 999, 1001, 1004 and 1006, we could take a false origin of 1000 and subtract it from all the values, giving −3, −1, +1, +4 and +6. The mean of these values is 1.4, so the mean of the original values is 1000 + 1.4 = 1001.4. The choice of the value of the false origin is arbitrary; in the example 1000 was chosen not only to make the calculation easier but also because it was near the centre of the values.

Occasionally we might want to find the overall mean for two data sets (A and B), where the mean and number of variates are known for each set.

In such cases the following formula should be used.

$$\bar{x} = \frac{n_A\bar{x}_A + n_B\bar{x}_B}{n_A + n_B}$$

Measures of variability and dispersion

While it is useful to describe a data set by an 'average' measure, there still remains the problem of describing the rest of the values. The data could spread out from the 'average' in quite different ways. There are, as with the measures of central tendency, various methods that can be used to measure this spread.

The range

This is simply the difference between the extreme values of a data set.

Example

Data set: 10, 12, 14, 16, 18, 20. Range is therefore $20 - 10 = 10$. This measure can obviously be affected by extreme values. If the data set was, for example, 2, 8, 9, 10, 11 and 20 then the range would be 18. For this reason it is not often used.

The inter-quartile range

To overcome this problem a measure is required which is not affected by extreme values. The simplest such measure is the inter-quartile range. You have already found that the median divides a data set into two equal halves. The quartiles divide the data set into four equal parts – the difference between the upper and lower quartiles being the inter-quartile range, i.e. between 25 and 75 per cent of the data set.

|← Inter-quartile range →|

Lower quartile **Median** **Upper quartile**

Example

Data set: 80, 90, 70, 100, 50, 40, 60

Firstly the data is arranged into a rank order.

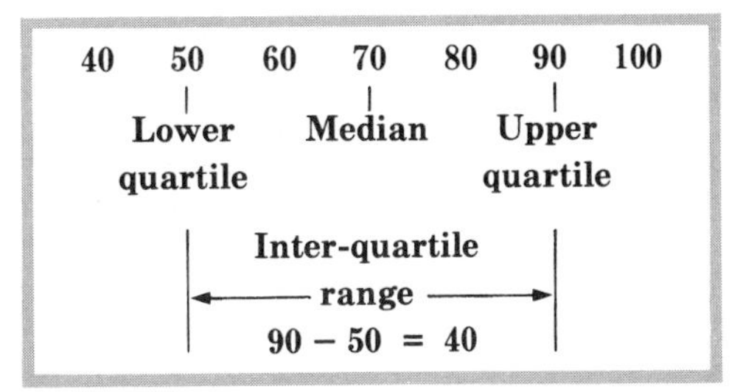

For grouped data it is better to draw a cumulative frequency curve in the same way as when estimating the median. The median is the value of x when f_{cum} is 50 per cent; the lower quartile the value of x when f_{cum} is 25 per cent and the upper quartile the value of x when f_{cum} is 75 per cent.

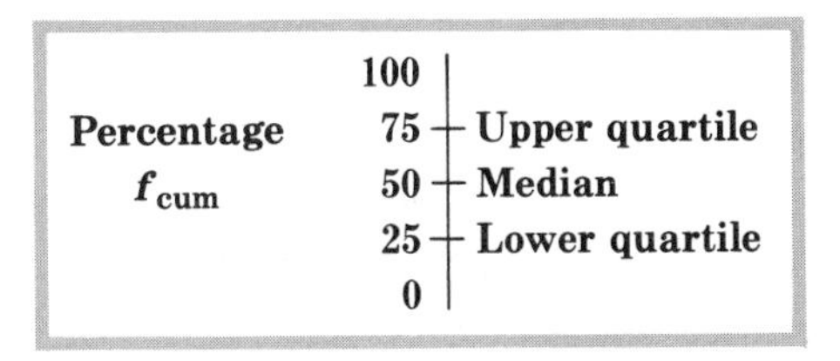

Let us look once again at the flower data from which the median was estimated using the probability grid. We can use the same technique to find the upper and lower quartiles and thus the inter-quartile range.

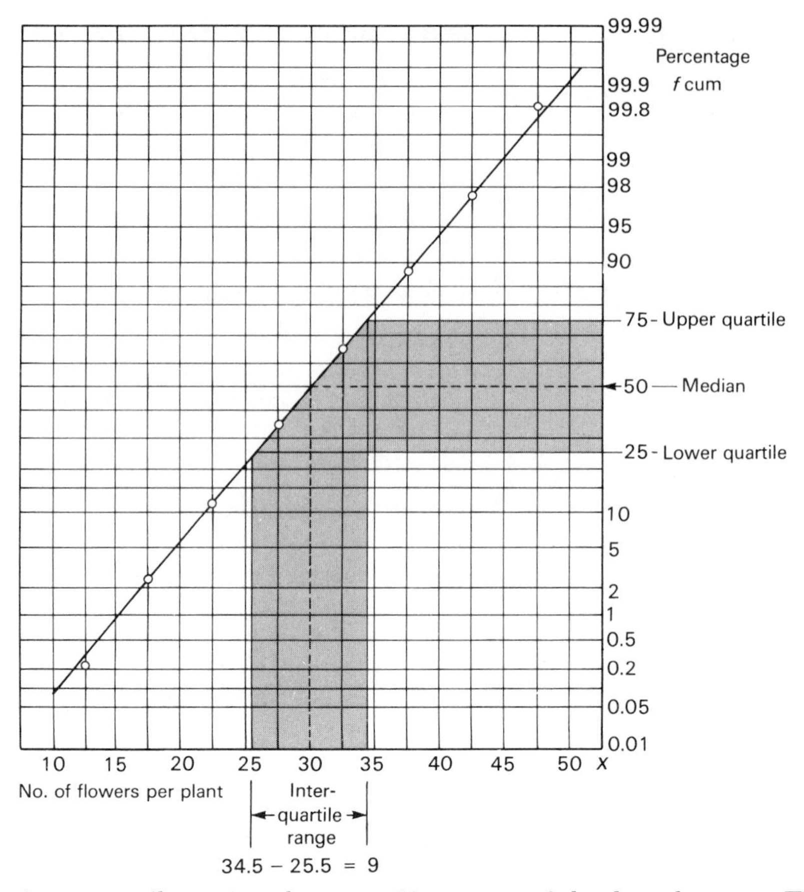

The inter-quartile range only covers 50 per cent of the data, however. This can be overcome by using ***inter-percentile*** ranges, e.g. 90 and 10 per cent, but the main problem with these measures is that they do not lend themselves to further mathematical treatment so their use is limited.

Variance and standard deviation

We still need to give greater definition to the variability found in biological data – the variance and standard deviation help in this. Where the inter-quartile range was related to the median, the variance and standard deviation are related to the mean.

The symbol for the variance of a population is $\hat{\sigma}^2$ and that for the standard deviation $\hat{\sigma}$ (small sigma). The same parameters for a sample are s^2 and s respectively. Since we are invariably dealing with data from samples the symbols s^2 and s are most often used.

For sample data the formula for the variance is

$$s^2 = \frac{\Sigma(x - \overline{x})^2}{n - 1}$$

You have come across the various symbols when you looked at the calculation of the mean. $(x - \overline{x})$ represents the distance or deviation from the mean of any value of x measured in the same units. $\Sigma\,(x - \overline{x})$ is the sum of all the deviations from the mean and should always add up to 0. If, however, we forget about the + and − signs then we have a rough estimate of the spread of the variates. This measure is known as the ***mean deviation***, but it is not normally used. Each of the deviations from the mean is squared $(x - \overline{x})^2$ (this gets rid of the + and − signs); then these squared deviations are summed: $\Sigma(x - \overline{x})^2$. The division by $(n - 1)$ is made because it has been shown that when deviations are taken from a sample mean instead of from a population mean, division by $(n - 1)$ gives a better estimate of the variance. The square root of the variance gives the value of the standard deviation.

Example

When calculating the variance always try to arrange things in a convenient and orderly manner so that errors are minimised.

x	$\bar{x}$	$(x-\bar{x})$	$(x-\bar{x})^2$
10	8	2	4
7	8	−1	1
6	8	−2	4
8	8	0	0
9	8	1	1

$\Sigma x = 40$ $\quad\Sigma(x-\bar{x}) = 0$ $\quad\Sigma(x-\bar{x})^2 = 10$

$n = 5$ $\quad$ ignoring

$\bar{x} = 8$ $\quad$ signs $= 6$

(mean deviation)

$$s^2 = \frac{\Sigma(x-\bar{x})^2}{n-1} = \frac{10}{4} = 2.5$$

$$s = \sqrt{2.5} = 1.58$$

Summary of the procedure

(i) Add up the values of x to give Σx.
(ii) Determine the value of n.
(iii) Divide Σx by n to give the value of $\bar{x}$.
(iv) Subtract the mean value from each value of x to give $(x-\bar{x})$.
(v) Add up the values of $(x-\bar{x})$ to give $\Sigma(x-\bar{x})$. This should equal 0.
(vi) Square the $(x-\bar{x})$ values to give $(x-\bar{x})^2$.
(vii) Add up the squared values of $(x-\bar{x})$ to give $\Sigma(x-\bar{x})^2$.
(viii) Divide this value by $n-1$ to give the variance s^2.
(ix) Determine the standard deviation s by taking the square root of the variance.

You can simplify matters by using a calculator with statistical functions. When determining the variance and standard deviation from grouped data the formula is $s^2 = \dfrac{\Sigma f(x-\bar{x})^2}{\Sigma f - 1}$

Using the flower data again:

x	f	fx	$\bar{x}$	$(x-\bar{x})$	$(x-\bar{x})^2$	$f(x-\bar{x})^2$
10	1	10	30	−20	400	400
15	10	150	30	−15	225	2250
20	40	800	30	−10	100	4000
25	100	2500	30	− 5	25	2500
30	135	4050	30	0	0	0
35	100	3500	30	5	25	2500
40	40	1600	30	10	100	4000
45	10	450	30	15	225	2250
50	1	50	30	20	400	400

$\Sigma f = 437$ $\quad\Sigma fx = 13\,110$ $\quad\Sigma(x-\bar{x}) = 0$ $\quad\Sigma f(x-\bar{x})^2 = 18\,300$

$$\bar{x} = \frac{\Sigma fx}{\Sigma f} = \frac{13\,110}{437} = 30 \qquad s^2 = \frac{\Sigma f(x-\bar{x})^2}{\Sigma f - 1} = \frac{18\,300}{436} = 41.97$$

$$s = \sqrt{41.97} = 6.48$$

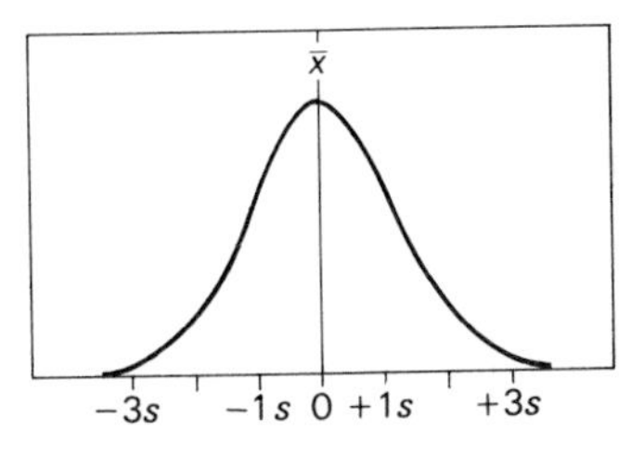

This distribution approximates a 'normal' distribution, as we found out when we plotted the percentage cumulative frequencies on the probability grid. Using such a technique we can estimate the standard deviation. Before we try this technique let us look at the relationship between the standard deviation and the theoretical model of the 'normal' distribution.

The theoretical model is based on the mean being 0 and the intervals on the x axis are marked in terms of the standard deviation, which is 1.

EXERCISE

1 Using again (you are probably getting 'fed-up' with these data by now) the flower data (x and f above), draw an accurate histogram on the grid provided on Work Sheet 49.

We have already computed the mean and standard deviation for these data: $\bar{x} = 30$ and $s = 6.48$.

2 On the grid draw vertical lines at values on the x axis corresponding to x (already drawn); $x \pm s$ (the mean plus one standard deviation and the mean minus one standard deviation; use a value of $s = 6.5$ to simplify the task); $x \pm 2s$ and $x \pm 3s$.

3 Count the total number of small squares (4 mm^2) enclosed by the histogram – note that the grid has been designed to simplify this task. Enter this value on Work Sheet 49.

4 Count the number of squares in the area bounded by the $x \pm s$ lines and express this number as a percentage of the total number of squares found in question 3. Enter this percentage on the Work Sheet.

5 Repeat this procedure for the number of squares bounded by the $x \pm 2s$ lines.

6 Repeat the procedure again for the number of squares bounded by the $x \pm 3s$ lines.

Range	Percentage
$\pm s$	68.27
$\pm 2s$	95.45
$\pm 3s$	99.73

For the purely theoretical model of the normal distribution, the percentages falling within the different ranges of standard deviation are as in the table.

7 Do your results for questions 4, 5 and 6 agree with these figures?

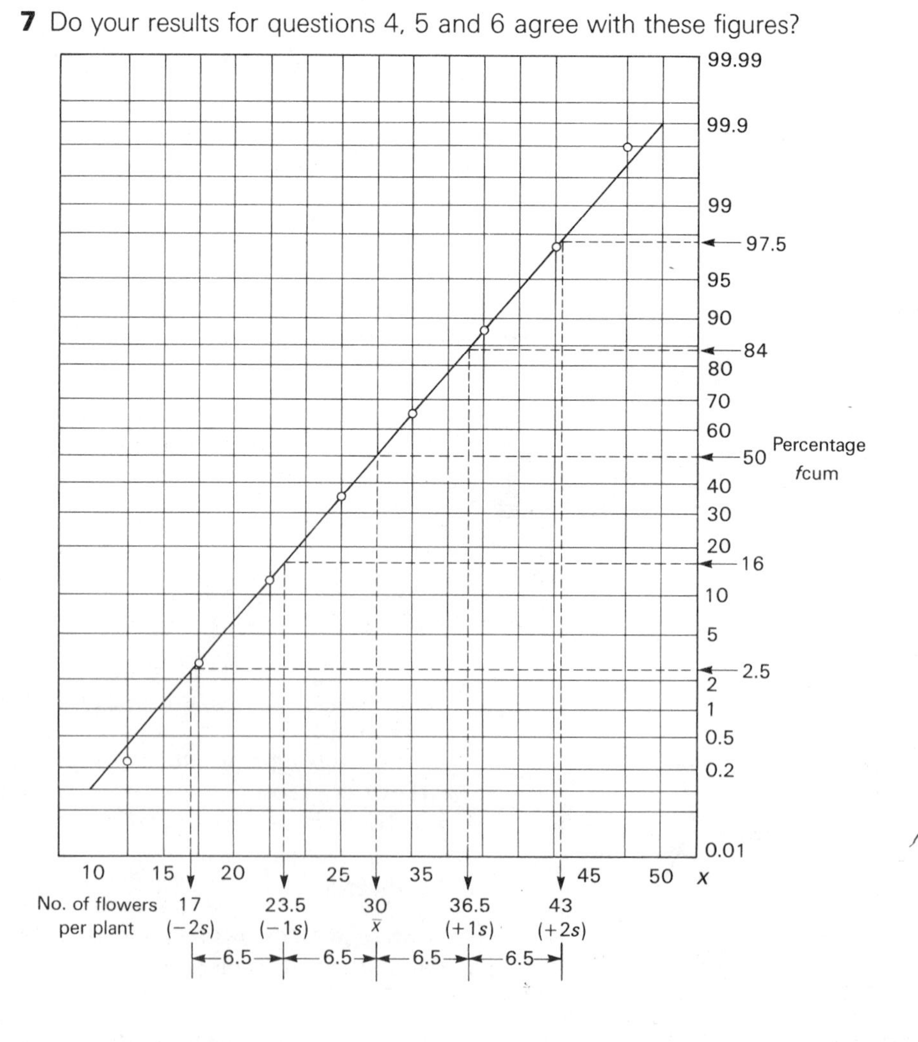

If a distribution approximates to a normal distribution, then the predictions which apply to the normal distribution can be applied to it also. Since the distribution of flowers was approximately normal, 68 per cent of the values should fall within the range plus and minus one standard deviation. From this information it should be possible to estimate the standard deviation from the percentage cumulative frequency graph that was drawn from the data. Dividing 68 per cent by 2, since the two halves of the graph are identical, the two values of percentage f_{cum} will be 50 per cent minus 34 per cent and 50 per cent plus 34 per cent, i.e. 16 per cent and 84 per cent respectively. When we project from the percentage f_{cum} values of 16 and 84 we find that the values on the x axis are 23.5 and 36.5 respectively. The value of 23.5 represents one standard deviation from the mean (known to be 30), and 36.5 also represents one standard deviation from the mean; the standard deviation is therefore $30 - 23.5 = 6.5$, or $36.5 - 30 = 6.5$. This figure compares very favourably with the computed value of 6.48.

Ninety-five per cent of the readings will fall within $\pm 2s$, i.e. between percentage f_{cum} 97.5 and 2.5. When these are projected we get x values of 43 and 17 respectively. (These lines are all shown on the graph.)

3.1.5 Variations on the normal distribution

The normal distribution can exhibit various forms depending on the values of the mean and standard deviation.

WORK SHEETS 50 and 51

EXERCISE

1 Draw histograms of the data given on Work Sheet 50 on the linear grid provided. Use different colours to *outline* each of the three distributions. Remember to label clearly each distribution.

2 Complete the table on Work Sheet 51, which has been started for you. Since the total number of specimens for each species is 100, the cumulative percentages will be equal to the cumulative numbers.

3 Draw line graphs of the data on the probability grid on Work Sheet 51. Remember that the points should be plotted to the right-hand side of the class (see p. 67).

4 Compare the slopes of the lines with the histograms you have drawn on Work Sheet 50. What conclusions can you come to?

5 *From the graph*, estimate the mean ($\bar{x}$) for (i) *H. vulgata*; (ii) *H. hypotheticus*; (iii) *H. maximus*. Compare your estimated values with the appearance of the histograms and comment.

6 *From the graph*, estimate the standard deviation (s) for each of the species. Compare the values with the appearance of the histograms and comment.

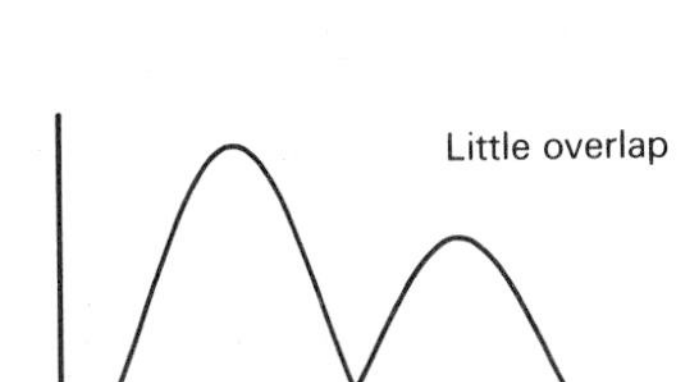

Bi-modality

Occasionally frequency distributions exhibit two 'humps' or modal classes. Such distributions are appropriately called ***bi-modal***. When two modal values are present there is a possibility that there are two different populations present, although of course they overlap; the amount of overlap can vary considerably.

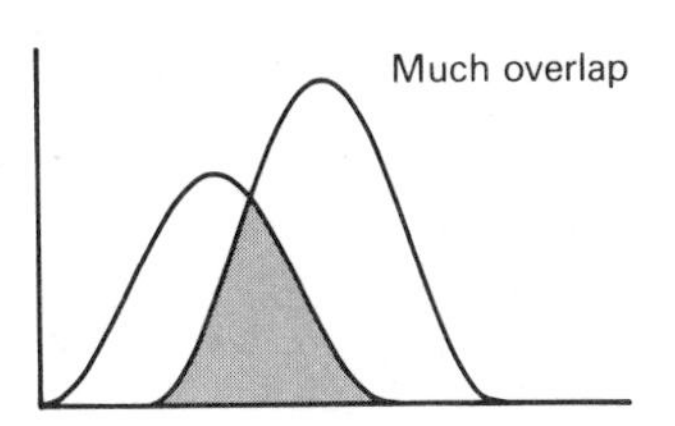

Example

Nuclella lapillus is a carnivorous gastropod commonly found on rocky shores. It feeds on barnacles, mussels, periwinkles and limpets. It deals with tough-shelled prey by drilling a neat round hole through the shell by means of a file like radula backed up by acid secretions.

Nucella lapillus

One day in September a sample of 300 *Nucella* was collected at random from a projecting headland. The heights of the shells were measured to the nearest 0.1 mm and the results displayed as a frequency distribution.

x Heights (mm)	f	f_{cum}	Percentage f_{cum}
10 < 11	3	3	1
11 < 12	6	9	3
12 < 13	12	21	7
13 < 14	24	45	15
14 < 15	30	75	25
15 < 16	27	102	34
16 < 17	18	120	40
17 < 18	18	138	46
18 < 19	24	162	54
19 < 20	30	192	64
20 < 21	36	228	76
21 < 22	30	258	86
22 < 23	21	279	93
23 < 24	12	291	97
24 < 25	6	297	99
25 < 26	3	300	100

When these data are graphed as a histogram (see graph below), the distribution is distinctly bi-modal. This suggests that there are possibly two populations present. We can speculate on what they might be, e.g. there could be two age groups, one distribution perhaps representing 1-year-old *Nucella* and the other distribution 2-year-olds. This hypothesis could be tested by trying to estimate the ages. Bi-modal distributions often suggest hypotheses which can be tested.

The two populations (A and B) overlap. By joining up the mid-points of the classes it is possible to outline the distributions more clearly (see graph).

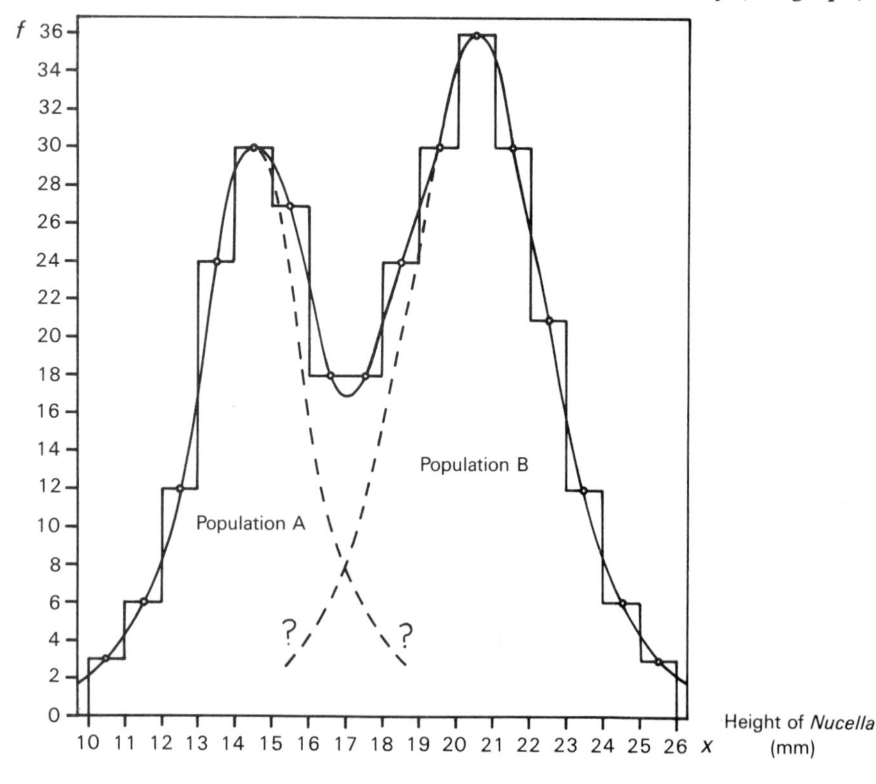

The dotted lines represent possible limits for each of the two populations. Is there any method by which the two populations could be separated more accurately?

The f values are cumulated (see column 3 in the table above), and these figures are then given as percentages (column 4, percentage f_{cum}). What would the graph be like if the cumulative percentages were plotted on a probability grid?

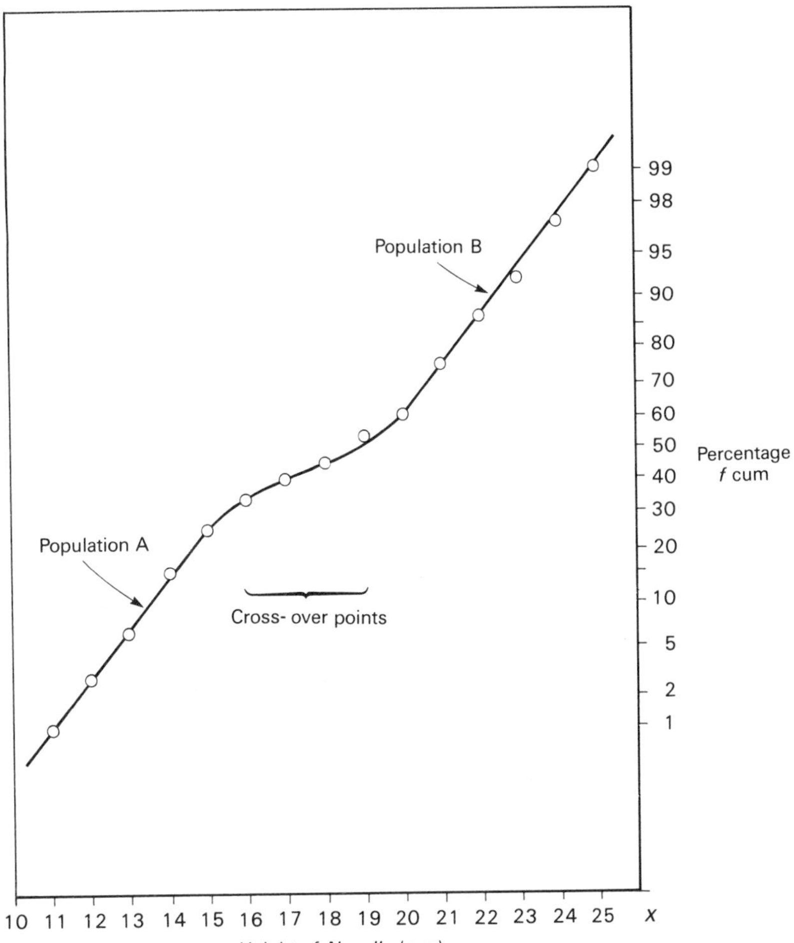

As you can see, there is a 'kink' in the straight line. This indicates a bi-modal distribution, but since the two ends of the line are straight the two contributary populations must be normal. The points where the cross-over from population A to population B occurs are 16, 17, 18 and 19.

The straight line representing population A is extrapolated, as is the line for population B (see the dashed lines on the graph on the next page). When the distance between the two population lines is measured (vertically) it is found to be 15 mm. The difference between the population A extrapolated line and the cross-over points indicates the contribution from population B, and the difference between the population B extrapolated line and the cross-over points indicates the contribution from population A.

When these differences are measured they are found to have the following values (mm).

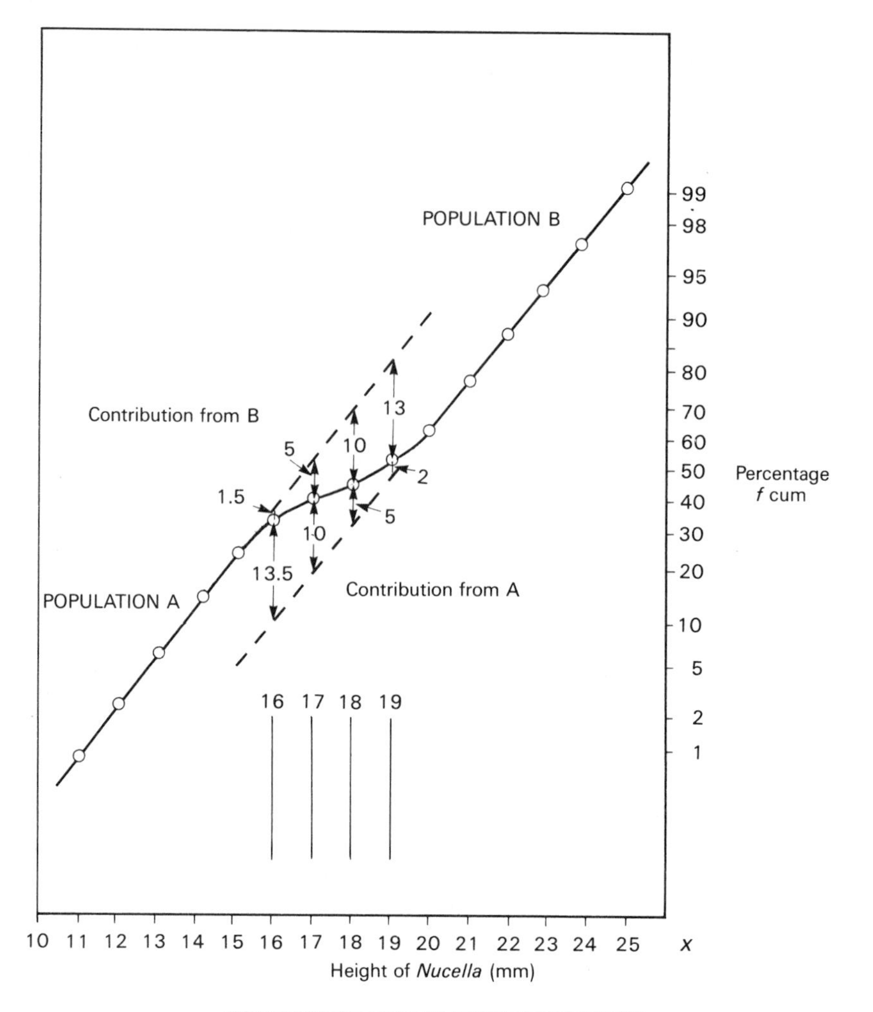

x	16	17	18	19
Pop. A	13.5	10	5	2
Pop. B	1.5	5	10	13

The total numbers for each of the x values 16, 17, 18 and 19 were 27, 18, 18 and 24 respectively, so the actual numbers making up each contribution should be in the same ratio as the measurements.

Point 16: 27 should be in the ratio 13.5 A : 1.5 B so A = 24 and B = 3;
Point 17: 18 should be in the ratio 10 A : 5 B so A = 12 and B = 6;
Point 18: 18 should be in the ratio 5 A : 10 B so A = 6 and B = 12;
Point 19: 24 should be in the ratio 2 A : 13 B so A = 3 and B = 21.

The actual frequencies of each distribution at the cross-over points have therefore been determined.

x	*f*A	*f*B
16	24	3
17	12	6
18	6	12
19	3	21

A histogram can now be drawn which shows the distributions of the two populations, A and B. This technique will only work satisfactorily when the standard deviations of the two populations are approximately equal i.e. when the two lines on the probability grid are parallel.

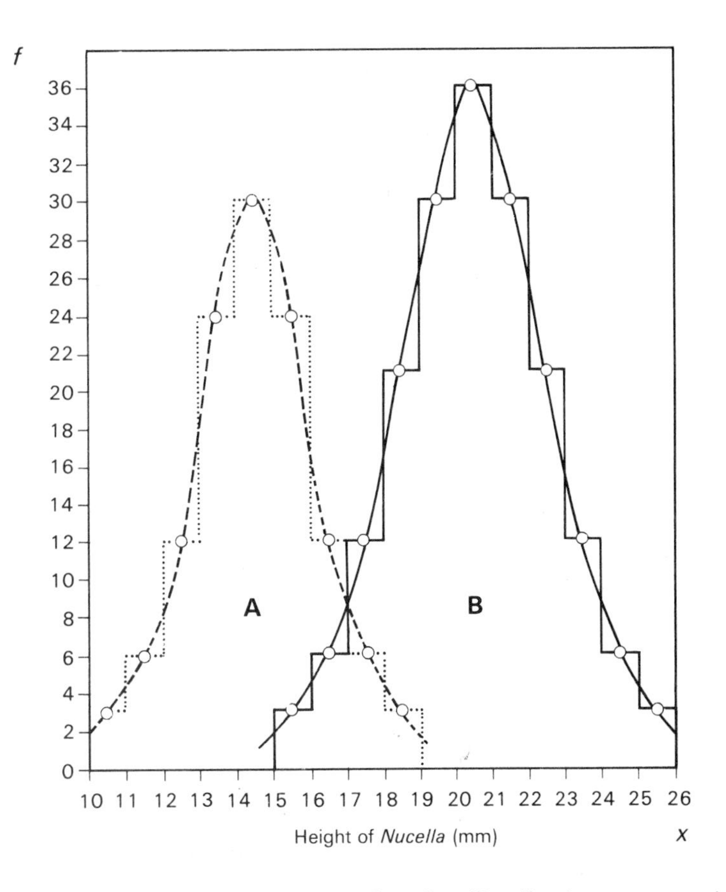

Taking the two populations separately, the distributions are as in the table.

x	Pop. A			Pop. B		
	f	f_{cum}	Percentage f_{cum}	f	f_{cum}	Percentage f_{cum}
10 < 11	3	3	2.5			
11 < 12	6	9	7.5			
12 < 13	12	21	17.5			
13 < 14	24	45	37.5			
14 < 15	30	75	62.5			
15 < 16	24	99	82.5	3	3	1.7
16 < 17	12	111	92.5	6	9	5.0
17 < 18	6	117	97.5	12	21	11.7
18 < 19	3	120	100.0	21	42	23.3
19 < 20				30	72	40.0
20 < 21				36	108	60.0
21 < 22				30	138	76.7
22 < 23				21	159	88.3
23 < 24				12	171	95.0
24 < 25				6	177	98.3
25 < 26				3	180	100.0

The means and standard deviations of each of the two populations can now be determined, either by computation from the data or by plotting the percentage cumulative frequencies on a probability grid. It is then possible to find out whether the two populations are statistically different from each other (see section 3.2).

Skew

Skewness is a deviation from the 'normal' curve where the distribution is asymmetrical, falling off more rapidly on one side than on the other. If the right-hand limb tapers off more gradually than the left-hand limb the distribution is said to be positively skewed or skewed to the left; if the left-hand limb is longer the distribution is negatively skewed or skewed to the right.

We have seen that when the normal distribution is symmetrical the mode, median and mean coincide. In skewed distributions they differ.

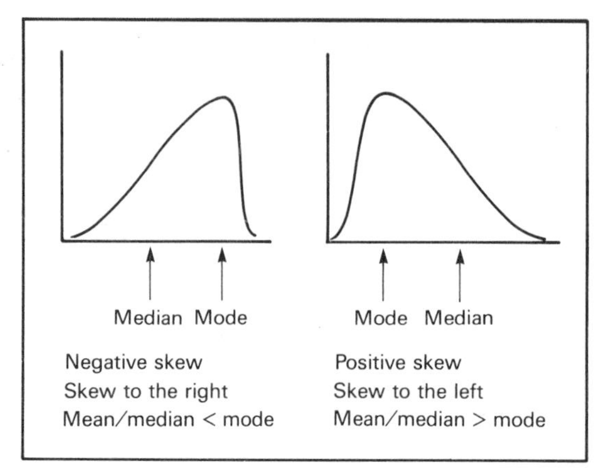

This difference between the mean and the mode can be used as a measure of skewness. The best parameter has been found to be the difference between the mean and the mode divided by the standard deviation:

$$S_k \text{ (coefficient of skewness)} = \frac{\text{mean} - \text{mode}}{s}$$

Biological distributions can exhibit every gradation from no skew to extreme skew. Whenever only a slight degree of skew is present, usually to the right, it can usually be ignored, but if the degree of skew is large then an explanation is required.

Occasionally a very extreme form of skew is found, where the first class has the highest frequency. This is known as a J-curve since the frequencies drop rapidly at first and then more slowly to zero.

Some distributions may be a combination of two strongly skewed distributions resulting in a U-shaped distribution, where there are peaks at both ends.

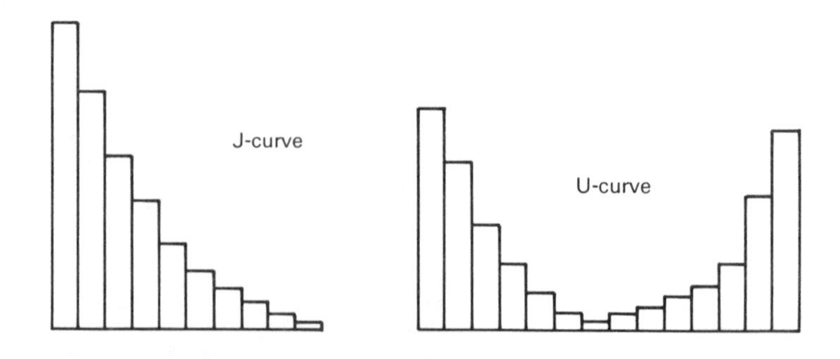

When both positively and negatively skewed distributions, drawn as histograms, are plotted on a probability grid, the positively skewed distribution curves upwards and the negatively skewed distribution curves downwards. The straight line is that for a symmetrical distribution exhibiting no skew.

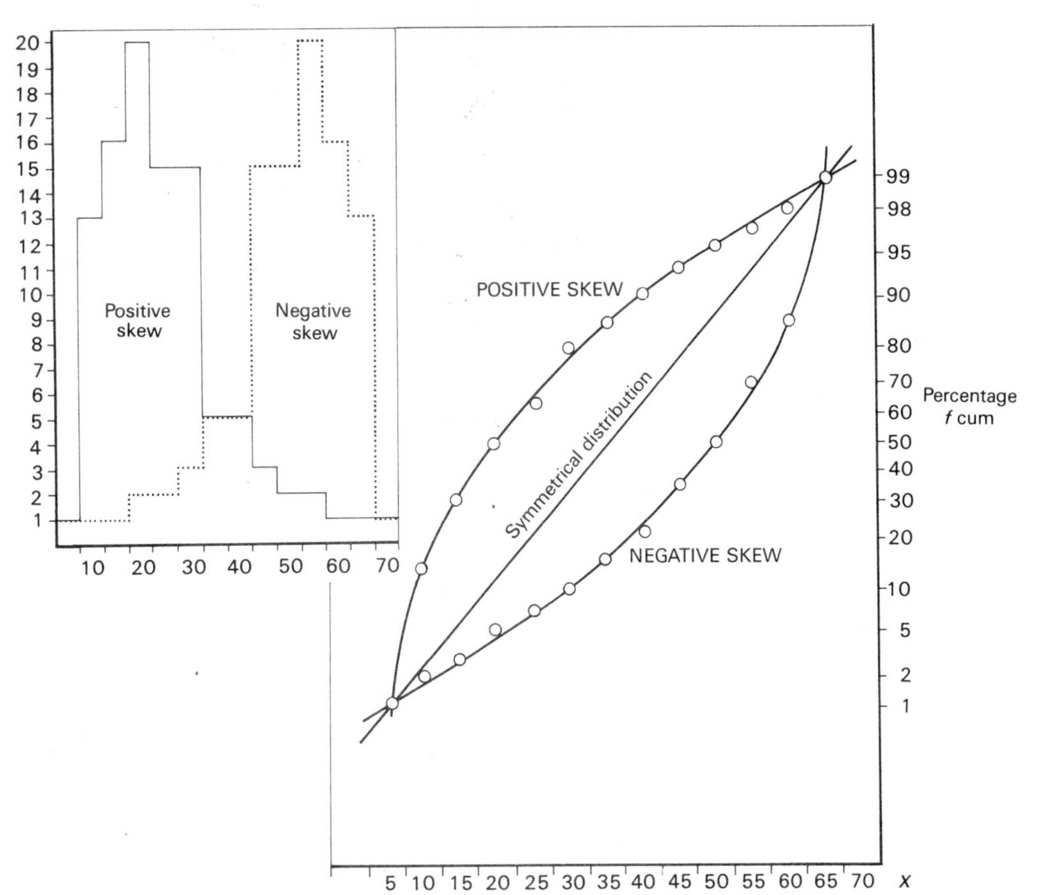

If the positively skewed distribution shown above is plotted on a logarithmic × probability grid, the result is a straight line. Such evidence can be of assistance in the analysis of population data.

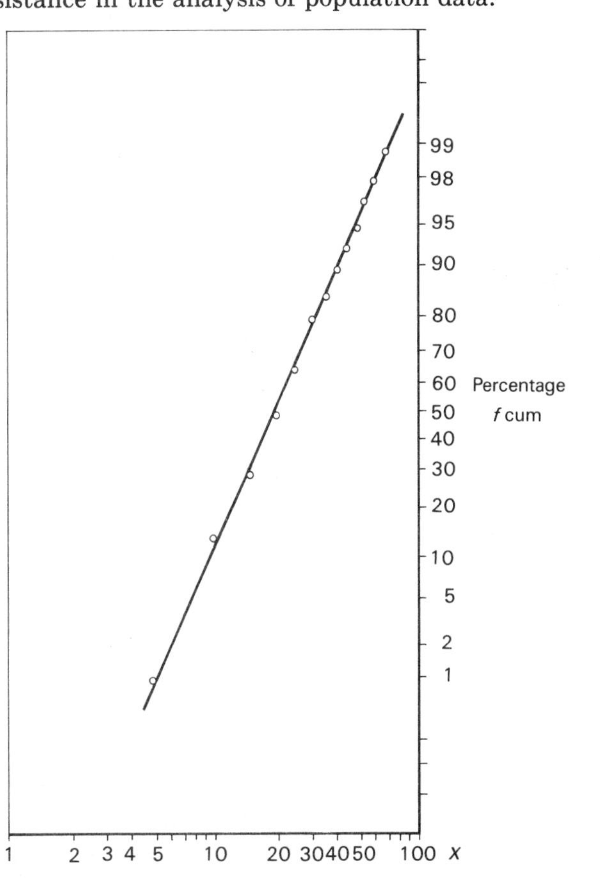

Kurtosis

The degree of spread about the mean is known as kurtosis (Greek: kurtos, meaning arched). It is an important characteristic since it gives some indication of the extent of the variability of a sample. Kurtosis could be described as meaning that the distribution is more pointed or flatter than the normal distribution.

If the distribution is distinctly peaked or sharper than a normal distribution it is called ***leptokurtic***, and shows low variability in the sample (Greek: leptos, meaning thin). If the distribution follows a normal distribution, or very close to it, it is known as a ***mesokurtic*** distribution. (Greek: misos, meaning middle, or in this case, intermediate). If the distribution is distinctly flattened relative to a normal distribution, it is called ***platykurtic***, and the sample shows a high degree of variability (Greek: platus, meaning flat).

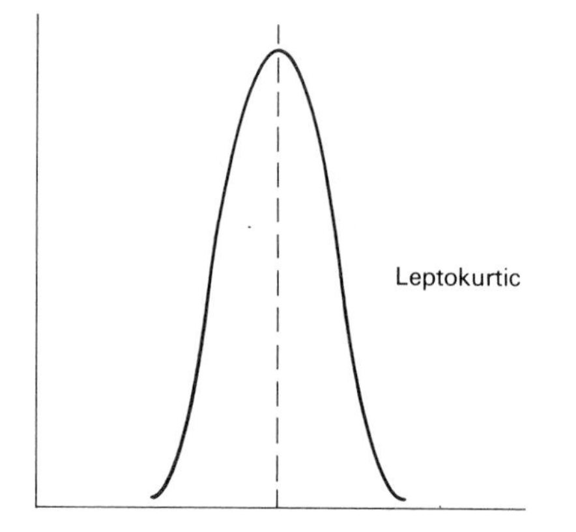

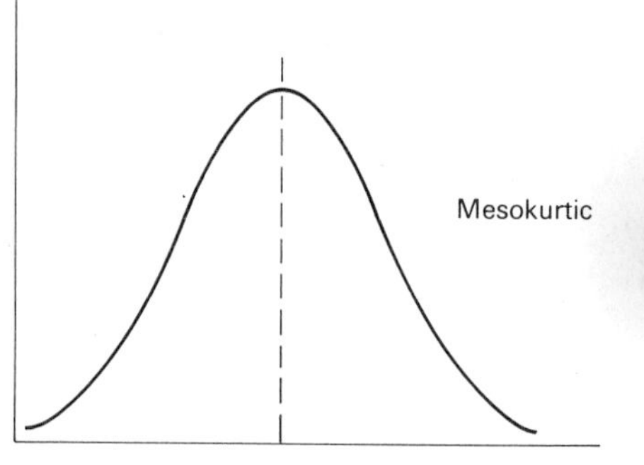

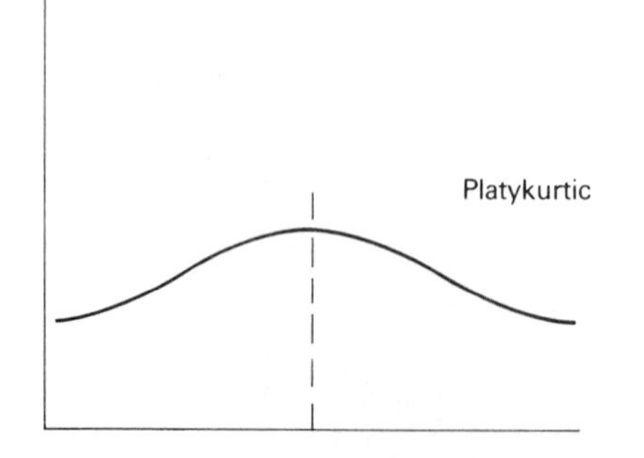

There is a coefficient of kurtosis (K_s) but its calculation is quite complicated. If $K_s = 0$ then the distribution follows the normal curve; if K_s is positive the distribution is leptokurtic, and if negative it is platykurtic.

3.2 PROBABILITY AND TESTS OF SIGNIFICANCE

In the previous section we found that the normal distribution curve was based on a theoretical model with a mean of 0 and a standard deviation of 1, the area under the curve representing the total population (100 per cent). The properties of the curve were such that within the range $\bar{x} \pm s$ the area under the curve is 68.27 per cent of the total; within $\bar{x} \pm 2s$ the area is 95.45 per cent and within $\bar{x} \pm 3s$ the area is 99.73 per cent.

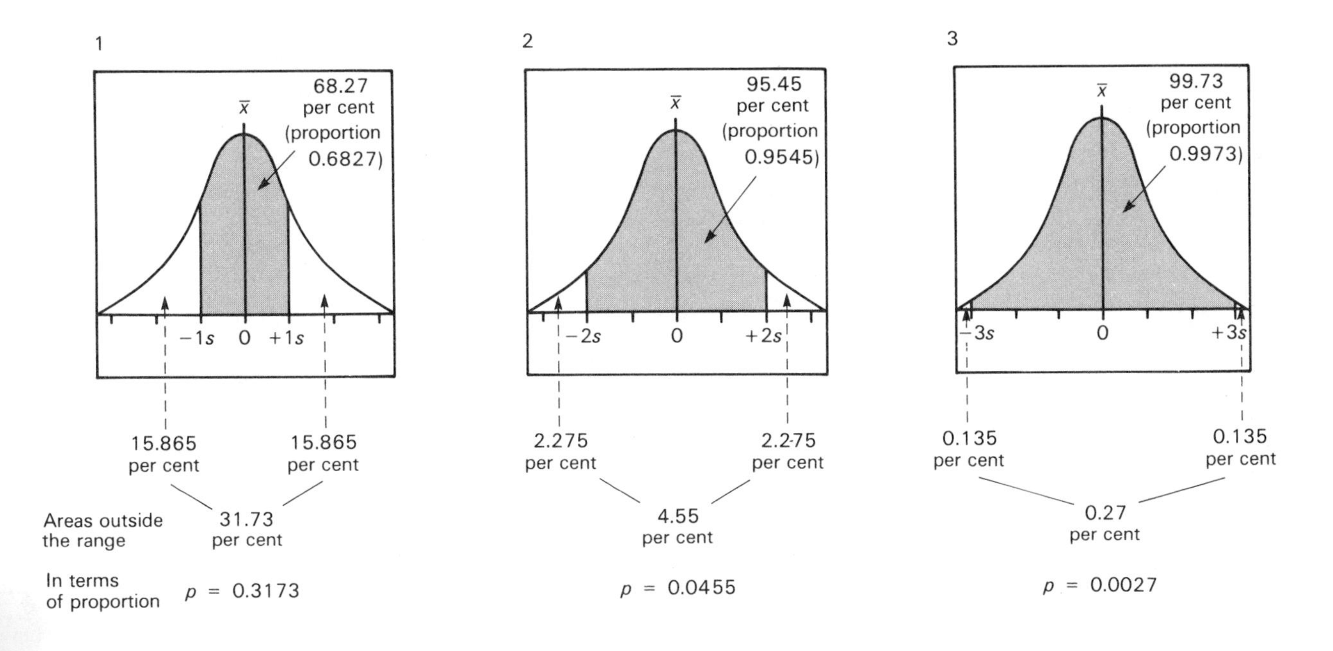

To find out how the proportions relate to the ranges of the standard deviation examine the diagrams of the normal curves below.

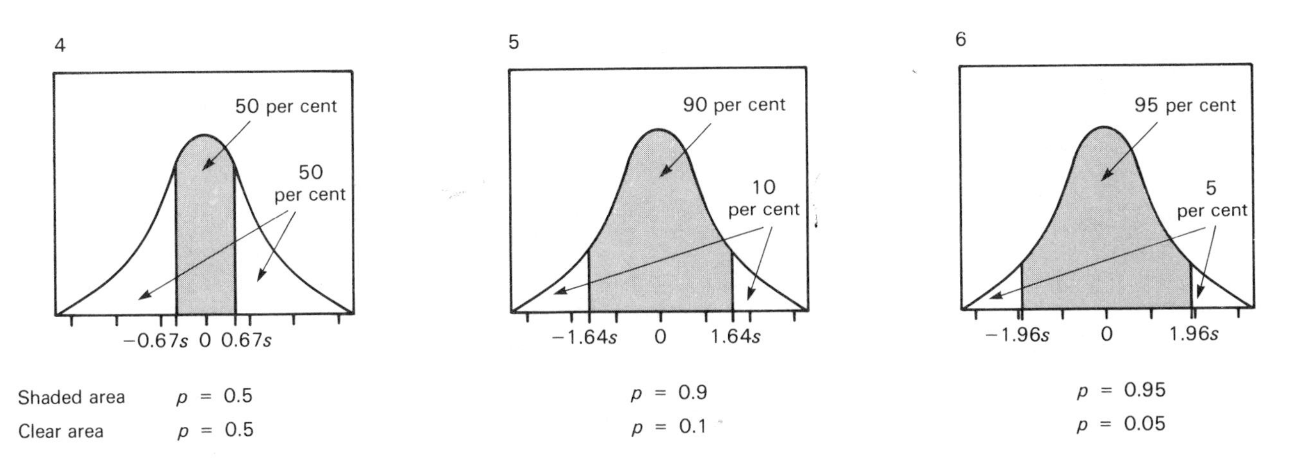

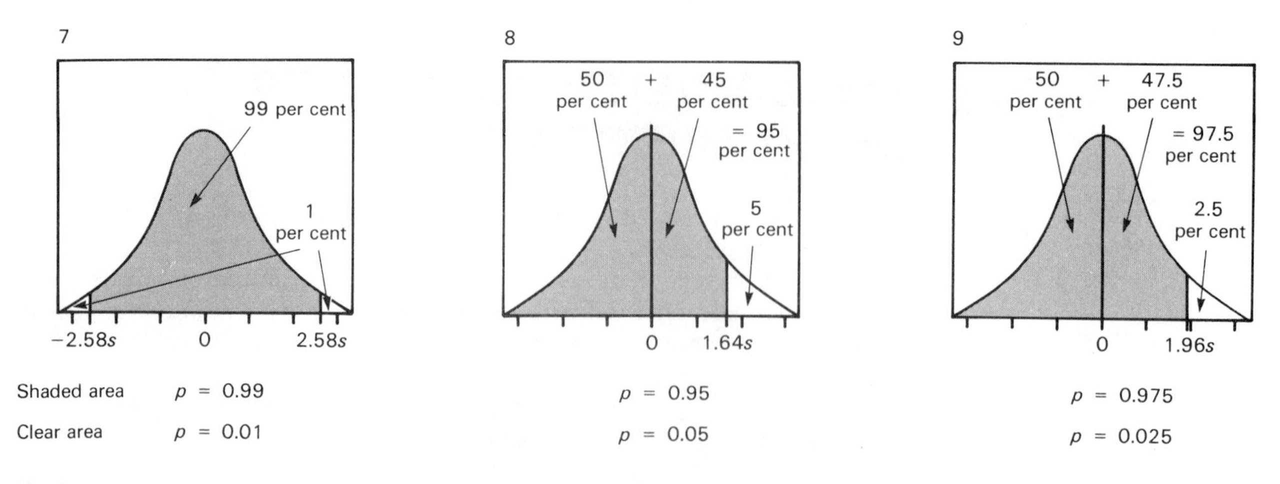

3.2.1 Use of the table of the standard normal distribution

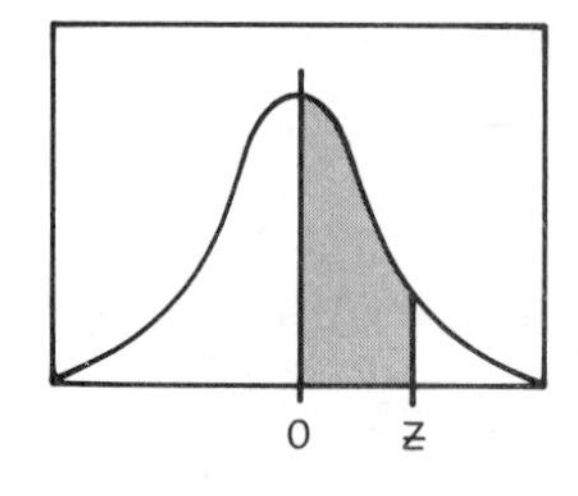

Accurate estimates of the areas, in terms of the proportion of the total area, are given in a table of the standard normal distribution. This table is given in Appendix 2. In this table areas are given for the complete range of units of standard deviation, which must be in the standardised form, i.e. with a mean of 0 and a standard deviation of 1. The proportions only include half of the curve, from the mean of 0 to any particular value Z (we will find out how to calculate this value later).

Examine again the diagrams in section 3.2 together with the table of areas of the standard normal distribution (Appendix 2, Page 115).

Diagram 1: range $-1s$ to $+1s$; proportion = 0.6827.
Table: when $Z = 1$ the proportion is 0.3413, but this applies only to the range 0 to 1 (see diagram above). The proportion 0 to -1 will be the same so $0.3413 + 0.3413$ (or 0.3413×2) = 0.6826. This proportion of the total area agrees with the diagram, allowing for a slight error due to rounding up.

Diagram 2: range $-2s$ to $+2s$; proportion = 0.9545.
Table: when $Z = 2$ the proportion is $0.4772 \times 2 = 0.9544$.

Diagram 3: range $-3s$ to $+3s$; proportion = 0.9973.
Table: when $Z = 3$ the proportion is $0.4987 \times 2 = 0.9974$.

EXERCISE

1 Carry out the same procedure for diagrams 4, 5, 6 and 7, checking the proportions against those determined from the table.

Diagram 8: proportion = 0.95
Table: $Z = 1.64$; proportion is 0.4495 *plus* 0.5 to include the complete other half of the area under the curve; so $0.4495 + 0.5 = 0.9495$.

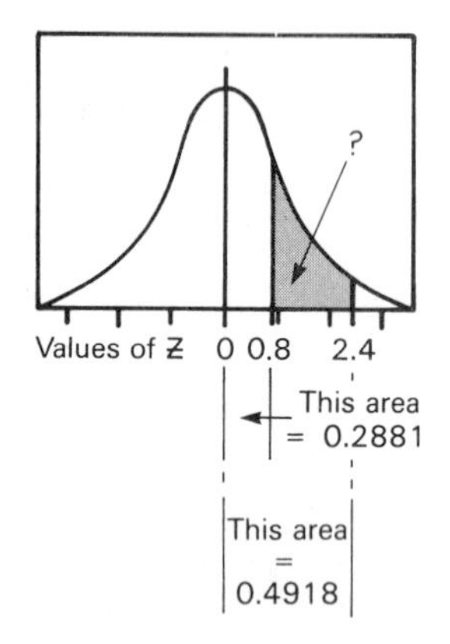

2 Carry out the same procedure for diagram 9 to find if the proportions agree.

Using the standard normal distribution it is possible to work out the proportion of the area between any two values of Z

Example

If the two Z values were 0.8 and 2.4 what would be the area? Looking at the diagram we see clearly the area to be determined. From the table we find that the area from $Z = 0$ to $Z = 0.8$ is 0.2881, and from $Z = 0$ to $Z = 2.4$ is 0.4918. The proportional area required is therefore the difference between these two values (see diagram), i.e. $0.4918 - 0.2881 = 0.2037$ (or 20.37 per cent).

What happens if one value of Z is negative and the other value positive?

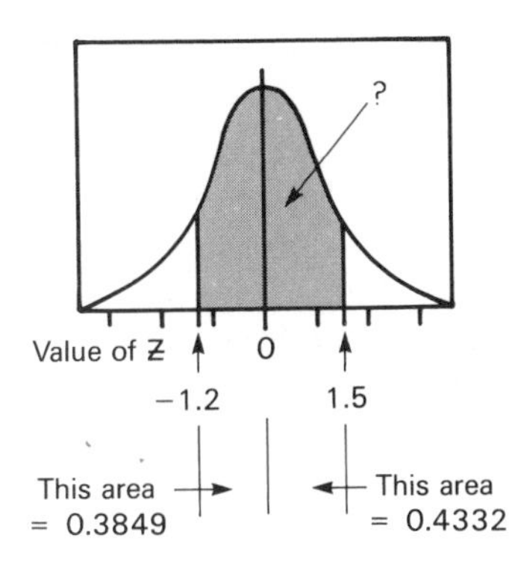

From the table, the area from $Z = 0$ to $Z = 1.2$ (since the curve is symmetrical forget about the minus sign) is 0.3849; the areas from $Z = 0$ to $Z = 1.5$ is 0.4332; the area required will therefore be the sum of these two areas (see diagram); $0.3849 + 0.4332 = 0.8181$ (or 81.81 per cent).

In order to be able to apply the standardised normal table using observational and experimental data it is necessary to convert the parameters ($\bar{x}$ and s) of a variate into standard form, i.e. into a value of Z.

The standardised normal deviate

Z is known as the standardised normal deviate and is computed using the formula

$$Z = \frac{x - \bar{x}}{s} \quad \text{(It is sometimes symbolised as 'd'.)}$$

?
Value of Z 0
1.81
0.4649
0.5

Example
Over the years records were kept of the litter sizes of a particular breed of pig. The mean litter size was found to be 10.2 with a standard deviation of 2.1. What percentage of litters could we expect to have 14 piglets or over?

First of all we work out the value of Z.

$$Z = \frac{x - \bar{x}}{s} = \frac{14 - 10.2}{2.1} = 1.81$$

From the table the area $Z = 0$ to $Z = 1.81$ is 0.4649. The total area to the right of $Z = 0$ is 0.5, so the area required is $0.5 - 0.4649 = 0.0351$ or 3.51 per cent. This is the percentage of litters which would be expected to have 14 piglets or over. The diagram illustrates the situation.

EXERCISES
What percentage of litters would be expected to have:

1 Eight piglets or less?

2 Nine piglets or more?

3 Between seven and twelve piglets?

4 Between five and nine piglets?

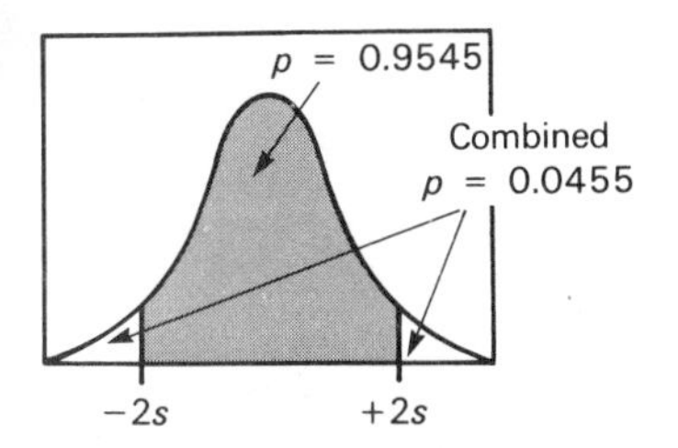

The areas above are given as proportions and percentages. They can also be given in terms of probability, where certainty is 1 and is equivalent to the total area under the curve. Particular values will then have levels of probability of falling within or outside certain limits (see diagram). These probability values have been developed from the theoretical normal curve which is based on an infinitely large population. In Biology we are nearly always dealing with samples from populations, whether we are gathering observational or experimental data. You must realise that the term population really means 'statistical' population, which defines the range over which conclusions can be drawn. It is not practicable to determine the parameters of *all* the blue-rayed limpets that exist or *all* the readings that could possibly be taken of a particular experiment! The standardised normal deviate Z was determined from the mean ($\bar{x}$) and standard deviation (s) of a sample, but it should really be determined from the mean (μ, pronounced mu) and the standard deviation ($\hat{\sigma}$, small sigma) of the population, i.e.

$$Z = \frac{x - \mu}{\hat{\sigma}}$$

Since μ and $\hat{\sigma}$ cannot be determined directly, $\bar{x}$ and s are used as estimates. It is logical, however, that the larger the sample the more closely will the values of $\bar{x}$ and s approach the values of μ and $\hat{\sigma}$ of the population – the size of the sample (n) must therefore be taken into account.

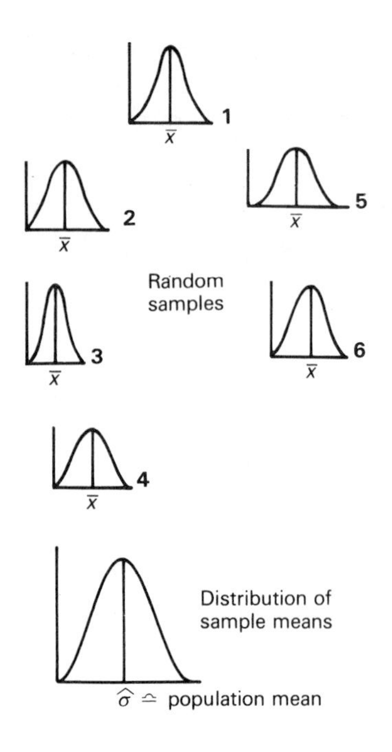

If we take a number of random samples from a population and compute the means for each sample, these means themselves will form a normal distribution. This distribution of sample means will also have a mean, which will be close to the population mean. We should also be able to estimate the standard deviation of the sample means from this distribution. Again, it would be quite impracticable or even impossible to obtain a large number of random samples, compute the means of all these samples and then determine the standard deviation of them. It has been found that this value can be determined from a single sample by dividing the standard deviation of a sample by the square root of the number of observations – this value is known as the ***standard deviation of the mean*** and is symbolised as $\hat{\sigma}_{\bar{x}}$. (This value is often called the standard error of the mean or simply the standard error, symbolised SE.)

$$\hat{\sigma}_{\bar{x}} = \frac{s}{\sqrt{n}} = \sqrt{\frac{s^2}{n}}$$

Confidence limits

From your knowledge of the 'normal' distribution you will remember that 95.45 per cent of the population lay between two standard deviations on either side of the mean. In Biology we normally refer to samples within 95 per cent. The limits within which the true value lies with a probability of 95 per cent are known as the ***95 per cent confidence limits***. The interval between the limits is known as the ***confidence interval***.

Such information can readily be added to a graph to enhance the amount of information presented. Usually graphs are plotted using only mean values; adding the confidence interval gives a visual indication of the variability of the data.

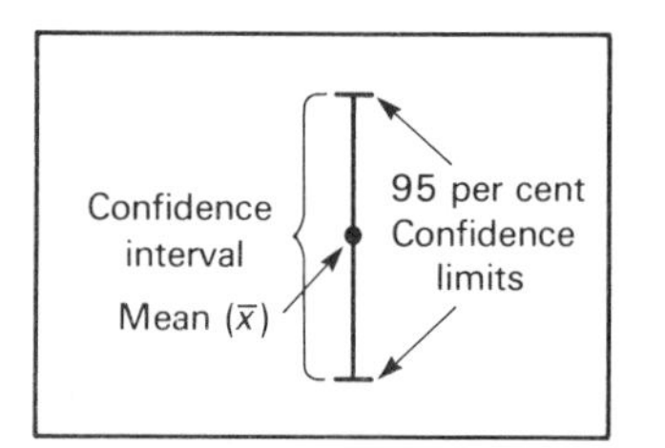

Determination of the confidence limits only requires that we know the mean and standard deviation of the sample, but the accuracy with which we can set these limits depends on the size of the sample. The larger the sample the more likely are its mean and standard deviation to approach those of the population from which it was taken.

The formula for calculating the 95 per cent confidence limits is

$$\bar{x} \pm t\hat{\sigma}_{\bar{x}} = \bar{x} \pm t \times \frac{s}{\sqrt{n}} \text{ or } \bar{x} \pm t \times \sqrt{\frac{s^2}{n}}$$

where $\bar{x}$ is the mean of the sample, s is the standard deviation of the sample, n is the number of observations and t is the value of t when $p = 0.05$ for any particular degrees of freedom $(n - 1)$.

The following table gives values of t for different degrees of freedom when $p = 0.05$. We will be returning shortly to t so don't concern yourself unduly about it at present. Regarding the 'degrees of freedom', just remember in this case that it is the number of observations minus one, i.e. $n - 1$.

Degrees of freedom (df)	Value of t	df	t	df	t	df	t
1	12.71	11	2.20	21	2.08	40	2.02
2	4.30	12	2.18	22	2.07	60	2.00
3	3.18	13	2.16	23	2.07	120	1.98
4	2.78	14	2.15	24	2.06	> 120	1.96
5	2.57	15	2.13	25	2.06		
6	2.45	16	2.12	26	2.06		
7	2.37	17	2.11	27	2.05		
8	2.31	18	2.10	28	2.05		
9	2.26	19	2.09	29	2.05		
10	2.23	20	2.09	30	2.04		

As you can see from the table, if the number of observations is greater than 10, the value of t is approximately 2: this makes the equation easier to remember: $\bar{x} \pm 2 \times s/\sqrt{n}$. You should also notice from the table that as the sample size increases the value of t decreases.

Example

In section 2.2.4 (p. 25) an experiment was described where yeast cell samples were placed on a haemocytometer slide, counts were made for each of five squares and from these mean values were obtained. These means were then used in plotting the graph. Although such points illustrate the general trend in growth of the yeast population, they do not indicate the spread of the observations around the means. This is overcome by adding the 95 per cent confidence limits to the plotted mean points.

From the data given on Work Sheet 14 the mean number of cells after 5 h was 124.4. The actual number of cells for each of the five squares was as follows.

Square number	1	2	3	4	5
Number of cells	110	105	157	130	120

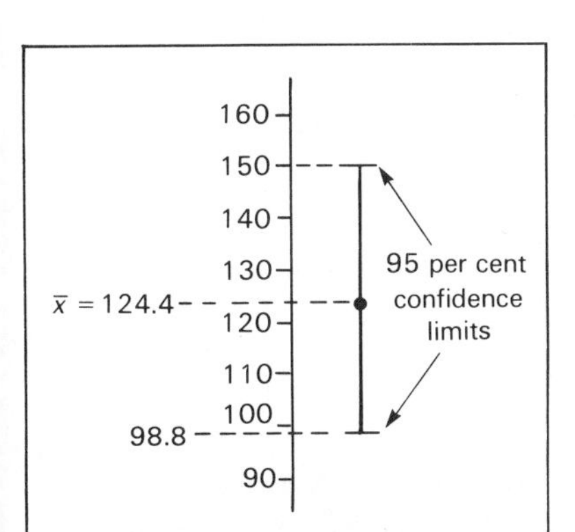

The total number of cells counted was 622 and so the mean number of cells at 5 h is $622/5 = 124.4$. The standard deviation when calculated = 20.6

Substituting in the equation $\bar{x} \pm t \times s/\sqrt{n}$, where $t = 2.78$ for 4 degrees of freedom ($n - 1$), we get

$$124.4 \pm 2.78 \times 20.6/\sqrt{5}$$

$$= 124.4 \pm 25.6$$

$$124.4 + 25.6 = 150$$

$$\text{and} \quad 124.4 - 25.6 = 98.8$$

When these 95 per cent confidence limits are added to the plotted mean they look like the diagram.

WORK SHEET 52

EXERCISE

Work Sheet 52 gives all the data for the experiment on the population growth of yeast.

1. Compute the standard deviations and enter them in the table provided.
2. Compute the 95 per cent confidence limits for each time and enter them in the table.
3. Arrange the scales on the grid. You must take into account the highest and the lowest confidence limits.
4. Plot the mean values as points on the grid.
5. Add the 95 per cent confidence limits to each of the plotted points, showing the confidence interval clearly.

We need only concern ourselves with the 95 per cent confidence limits since these are the ones normally used in Biology. Confidence limits can, however, be determined for any particular probability level, e.g. 90 or 99 per cent, by using the value of t for the appropriate p value, as given in the table of t – Appendix 3A, Page 116.

Example

Confidence limits can also be added to bar-graphs. In section 2.2.6 (p. 32), an experiment was described concerning the release of bubbles from *Elodea* under different colours of light. The results given on Work Sheet 24 are mean values from 10 replicates.

The actual readings for violet were as follows.

54	62	46	66	70	58	54	50	62	58

$(n = 10)$

The mean for these values is 58 and the standard deviation 7.3. Looking up the value of t from the table for 9 degrees of freedom gives us 2.26.

The 95 per cent confidence limits are therefore

$$58 \pm 2.26 \times 7.3/\sqrt{10} = 58 \pm 5.2$$

$$58 + 5.2 = 63.2 \quad \text{and} \quad 58 - 5.2 = 52.8$$

The limits and interval are added to the bar as shown below.

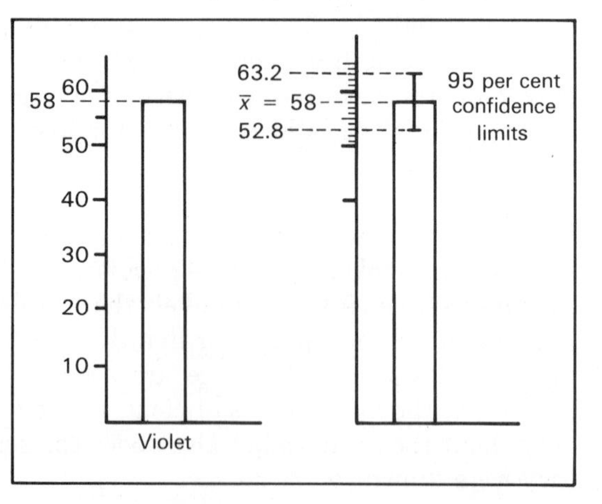

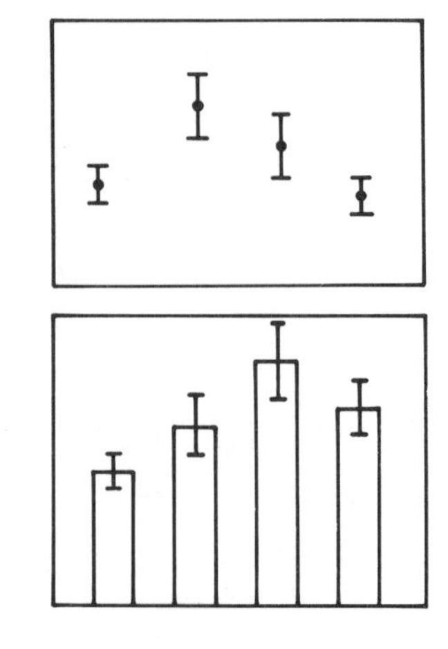

WORK SHEET 53

EXERCISE

Work Sheet 53 contains all the data from the experiment, giving the values of each of the ten replicates for each colour of light.

1. Compute the means and standard deviations for the other colours of light; enter these in the table provided.

2 Compute the 95 per cent confidence limits for each colour of light; enter these in the table provided.

3 Draw a bar graph of the data on the grid provided and add the 95 per cent confidence limits.

4 Which colour exhibited the greatest variation in the results, and which the least variation?

Nowadays many of the results published in research papers in graphical form have confidence limits plotted for each sample mean. Clearly this practice is very useful and the reader can see the variability of the data at a glance. Unfortunately some biologists use the confidence limits wrongly. They attempt to compare samples simply by inferring that:

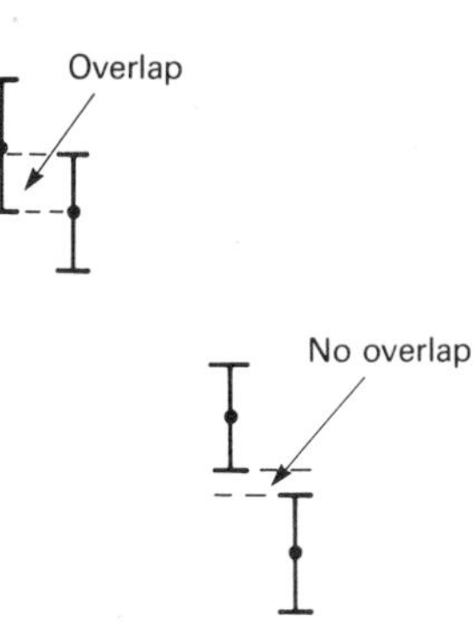

if the confidence limits overlap then there is no significant difference between the samples; or

if the confidence limits don't overlap then the samples are significantly different.

Such conclusions are incorrect. If you want to compare samples then the appropriate significance test should be used (see section 3.2.2, below).

3.2.2 Tests of significance

The main reason for carrying out observations and experiments in Biology is to test an idea or hypothesis; this idea could have arisen either through thinking about a problem or as the result of previous observations or experiments. Biology, as a science, is an on-going subject, answers to one question usually leading to further questions. Such work results in data sets, as you have seen throughout this text. What we want to know is whether the data supports the initial hypothesis or not; does the data enable us to accept our hypothesis as a valid explanation or do we have to modify the hypothesis, or even reject it?

We should by now appreciate that biological data is variable; it consists of small samples taken from large populations. Even though these samples are randomly chosen it is very unlikely that the data will conform exactly to the hypothesis, no matter how brilliantly it is thought out. To overcome this problem, tests of significance are used to determine whether the data conform to the hypothesis sufficiently closely for a decision to be made, not only about the samples but also about the general populations from which they were drawn.

Often the hypothesis requires a comparison between two or more samples. The degree of difference between samples may be very small, very large, or anything in between. The real purpose of the exercise is to find out, from the difference between the samples, whether they were drawn from the same (statistical) population or from different (statistical) populations. We cannot be certain about any statistical conclusion – a statistical test doesn't prove anything. It does, however, give us some degree of confidence in our conclusions.

Statistical tests of significance often cause the Biology student (and others too) a great deal of difficulty. At A-level standard you really only need to know about three of the tests – the t test, the 'chi-squared' test and the 'correlation coefficient' test. These three will suffice for most of the work, exam questions and project work that you will encounter.

The null hypothesis

Much of the difficulty in understanding significance tests arises from the use of the 'null hypothesis'. This is in effect a negative hypothesis: for example, if we were comparing two samples, the null hypothesis would be stated in the form 'there is no difference between the two samples'. It assumes that any differences that do occur are due to chance. The significance test then confirms whether to accept or reject the null hypothesis. If it is accepted then there is no difference between the two samples; if rejected there is a difference between them. This use of a sort of double negative leads to confusion; it is probably a legacy from the origins of statistical tests when they were used in quality control in industry (they still are); batches of manufactured goods were either accepted or rejected according to whether they were up to standard. The use of the null hypothesis should become clear as you work through the examples and problems.

Although most statistical texts still use the null hypothesis, it has been superseded in biological research by a much more straightforward procedure where the null hypothesis is not normally stated. It does seem a bit strange that if we carry out an experiment the whole purpose and motivation of which assumes that there is likely to be a difference between the experimental and control groups, that we should state that there is no difference and then go on to reject the statement. It is better just to find out whether a difference occurs or not.

The only way of overcoming this problem is to give both methods described below – you can then choose whichever one suits you best. You should however be aware of both methods.

Two main situations arise in biological work:

(i) Where the main (not the null) hypothesis is supported by a difference between the samples. Examples are experiments involving a treatment of some sort or samples from different localities.

(ii) Where the main hypothesis is supported by no difference between the samples. Examples of this type occur when results are predicted according to the hypothesis and the observed results are then compared with the prediction.

Since significance tests compare samples with a theoretical statistical model which is based on an infinitely large population with a normal distribution, it follows that the samples should be of a similar nature so that like can be compared with like. The larger the samples the more valid is the test. If only small samples are used then we must be cautious about any conclusions we arrive at. Replicating an experiment a number of times or taking more samples can overcome this problem. Both tests, but particularly the t test, also require that the variables under consideration are from normally distributed populations. Although many distributions will be binomial or

William S. Gosset

Poisson (see pp. 64, 65) the model of the normal distribution can in most cases apply, thus simplifying the matter.

The Student's *t* test

The statistic *t* was first published in 1908 by William S. Gosset, who had joined Guinness Brewery in 1899 from Oxford University as a scientist brewer – he was not a mathematician. At the time it was necessary to assess the potential of new barley varieties. It was not possible to rely on large volumes of data because the number of experimental plots was limited and, because of the growing season, only one experiment could be undertaken each year. Gosset observed that the mean and standard deviation of a large set of samples could be very accurately determined using the statistical techniques available at the time; for smaller sets of samples, typically below 30, the error progressively increased. It was thus necessary to take this increased error into account when drawing conclusions from data. The *t* statistic was devised to overcome this problem. From his observations Gosset was able to put numerical values to the variation associated with small sample sizes.

At the time the work was published, the Guinness Board did not allow employees to publish under their own names but they suggested either the pseudonyms 'Pupil' or 'Student'; Gosset chose 'Student' and so the statistic became known as Student's *t*. This statistic is nowadays used very widely in the analysis of data from natural and applied scientific research, medicine, market research, etc.

How can we compare samples statistically?

First we have to compute the standard deviation of the difference between the means of the two samples. In the previous section (p. 84) we found that the standard deviation of the mean was $s/\sqrt{n}$; the ***variance of the mean*** (the square of the standard deviation of the mean) will therefore be

$$\hat{\sigma}_{\bar{x}}^2 = \left(\frac{s}{\sqrt{n}}\right)^2 = \frac{s^2}{n}$$

If we want to compare two samples, A and B, then the variance of the difference of sample means will be the variance of the mean of sample A plus the variance of the mean of sample B, i.e.

$$\hat{\sigma}^2_{(\bar{x}_A - \bar{x}_B)} = \hat{\sigma}^2_{\bar{x}_A} + \hat{\sigma}^2_{\bar{x}_B} = \frac{s^2_A}{n_A} + \frac{s^2_B}{n_B}$$

Converting back to the standard deviation of the difference this will take the form

$$\hat{\sigma}_{(\bar{x}_A - \bar{x}_B)} = \sqrt{\hat{\sigma}^2_{\bar{x}_A} + \sigma^2_{\bar{x}_B}} = \sqrt{\frac{s^2_A}{n_A} + \frac{s^2_B}{n_B}}$$

Providing that the data are normally distributed the standard deviation between sample A and sample B will have the same characteristics with regard to probability as the normal distribution curve. If the probability of the ***actual*** difference between the means of samples A and B being greater than twice the standard deviation of the difference is nearly 0.05 (or, to put it another way, if the actual difference between $\bar{x}_A$ and $\bar{x}_B$ is greater than twice the standard deviation of the difference), then it is unlikely that a difference of this magnitude between the two sample means occurred by chance – some other factor must be involved.

Example

A parasitologist wanted to compare the number of lymphocytes present in the blood of patients infected by a blood parasite with the numbers in uninfected individuals. To save time and resources he limited the investigation to five individuals in each group.

Using prepared blood films, the lymphocyte counts obtained from equal areas of blood were as below.

Group A (infected patients)	150, 155, 152, 146, 152
Group B (uninfected individuals)	165, 170, 151, 164, 160

[NI]

Are these two groups statistically different?

The means and standard deviations were computed as below.

	Mean	Standard deviation
Group A	151	3.32
Group B	162	7.11

$$\hat{\sigma}_{(\bar{x}_A - \bar{x}_B)} = \sqrt{\frac{s^2_A}{n_A} + \frac{s^2_B}{n_B}} = \sqrt{\frac{(3.32)^2}{5} + \frac{(7.11)^2}{5}} = \sqrt{\frac{11.02}{5} + \frac{50.55}{5}}$$

$$= \sqrt{12.314} = 3.51$$

Now this standard deviation of the difference is compared with the ***actual*** difference between the means of samples A and B:

$\bar{x}_A - \bar{x}_B = 151 - 162 = 11$ (forget the sign – this is an absolute number)

The actual difference between the means (11) is just more than three times greater than the standard deviation of the difference (3.51). This means that the difference is more than three standard deviations in terms of the 'normal' curve. The probability of such a situation arising is only 0.0027 – the difference between the samples would only occur by chance 2.7 times in every 1000 – a very unlikely event! We can therefore conclude that there is a difference between the two samples. Looking back at the mean values we can state that the lymphocyte count in uninfected individuals (sample B : mean 162) was higher than in infected patients (sample A : mean 151).

This method only gives us rough estimates – if the difference between the means is two or three times the standard deviation of the difference. The Student's t test allows us to find out exactly how many times greater the actual difference between the sample means is than the standard deviation of the difference.

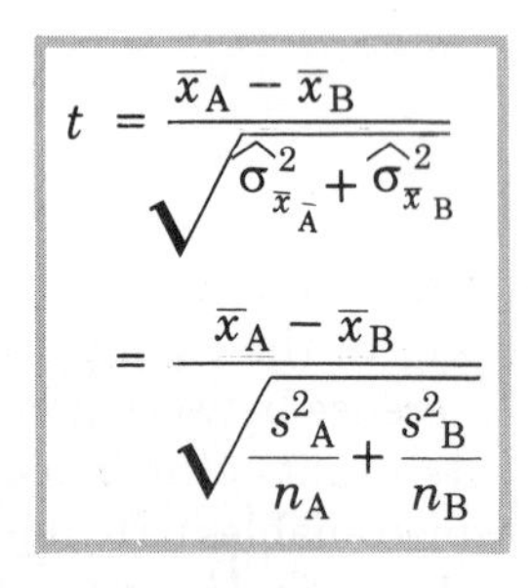

$$t = \frac{\bar{x}_A - \bar{x}_B}{\sqrt{\hat{\sigma}^2_{\bar{x}_A} + \hat{\sigma}^2_{\bar{x}_B}}} = \frac{\bar{x}_A - \bar{x}_B}{\sqrt{\frac{s^2_A}{n_A} + \frac{s^2_B}{n_B}}}$$

t is the actual difference between the means of the samples divided by the standard deviation of the difference between the means.

Examine the table of t distribution given in Appendix 3A. It differs from the one given in the previous section (p. 85) in that it gives the values of t for a range of values of p and not just for $p = 0.05$. You will see that down the left-hand side of the table are given ***degrees of freedom*** (df 1 to infinity) and along the top of the table are the p values (note: these ***decrease*** as you go to the right).

The table is entered at a particular number of degrees of freedom which is determined by

$(n_A - 1) + (n_B - 1)$ or $(n_A + n_B) - 2$, i.e. two less than $n_A + n_B$

EXERCISE

1 If sample A had 20 observations and sample B had 28 observations, how many degrees of freedom would there be?

We now move across the table of the t distribution (Appendix 3A) until a value of t is found which is as close as possible to the computed value of t, and the appropriate value of p is determined.

Example

If the value of t was 3.17 and there were 10 degrees of freedom, what would be the value of p?

Looking up the table of t (Appendix 3A) we go down the df column on the left of the table until we come to 10. Then we move to the right across the t value columns until we come to one with a value equal to that of the computed t value – the third column is 3.17. At the top of the table the p value for this column is 0.01 – the value of p is therefore 0.01. When talking about probability values totality is 1; they can be expressed as a percentage when totality is 100, i.e. in this case 1 per cent.

You might find using the graph (Appendix 3B) easier than using the table. Move along the X axis until you find the df = 10 line. Move up this line until you come to the intersection of the $t = 3.17$ line – this intersection coincides with the $p = 0.01$ curved line so the value of p is 0.01.

If the value of t is 3.1 and there are 20 degrees of freedom you will find that p lies between 0.01 ($t = 2.85$) and 0.001 ($t = 3.85$).

		***p* values**			
		0.10	**0.05**	**0.01**	**0.001**
d.f.	**20**	**1.72**	**2.09**	**2.85**	**3.85**

3.1 lies in here

The p value in such cases is described as being less than 0.01 (< 0.01). ***Always remember that the p values decrease as you move to the right.*** This can be a confusing point since we are normally used to values which increase as we move to the right.

EXERCISE

Try these questions, using both the table and the graph of t.

1 If $t = 2.94$ and df = 6 what is the value of p?

2 If $t = 4.57$ and df = 30 what is the value of p?

What do these values of p mean?

Using the null hypothesis

As we have already found out, we first make the assumption that any differences between the samples are due to chance. In the example given above the null hypothesis would be 'there is no difference between groups A and B' or 'there is no difference in the lymphocyte counts of infected and uninfected individuals'. The value of p that is obtained enables us to reject or accept this null hypothesis. ***The level of rejection in biological investigations is normally taken as p = 0.05 or the 5 per cent level.***

Look again at the table of t (Appendix 3A). We can see that the values of t get larger as we move to the right – the differences between the samples get larger. Any value of t to the ***left*** of the $p = 0.05$ column is considered small enough ***to accept*** the null hypothesis, i.e. there are no true differences between the samples, any differences being due the chance. Any value to the ***right*** of the $p = 0.05$ column is large enough to ***reject*** the null hypothesis, i.e. there is a difference between the samples. The further we move to the right (and the smaller the value of p) the more strongly the null hypothesis is rejected and there are conventions for the level of rejection:

if $p \leqslant 0.05$ the rejection can be considered ***significant***;
if $p \leqslant 0.01$ the rejection can be considered ***highly significant***;
if $p \leqslant 0.001$ the rejection can be considered ***very highly significant***.

If the rejection is found to be ***significant*** then we can be ***fairly confident*** in stating that the null hypothesis can be rejected and there is a difference between the two samples; if ***highly significant*** then we can be ***very confident*** in our statement and if ***very highly significant*** then we can be ***almost certain*** that the null hypothesis can be rejected.

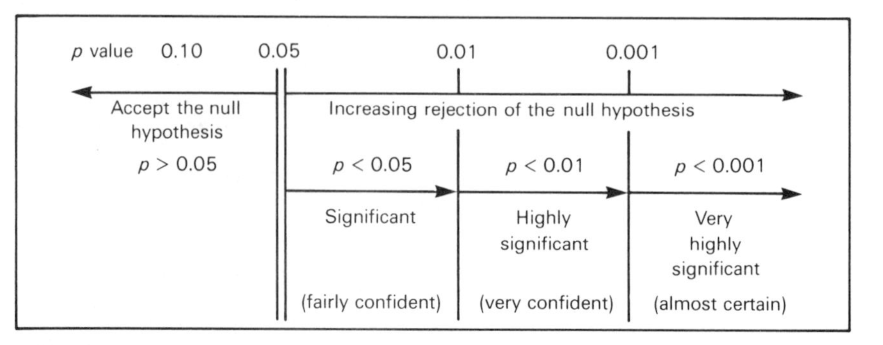

Not using the null hypothesis

The same technique is used except that the null hypothesis is dispensed with. Any p value greater than 0.05 (> 0.05) means that the differences between the samples are not significant (NS), any differences that do occur being due to chance. A p value of less than 0.05 (< 0.05) means that there is a difference between the samples. From $p = 0.05$ to $p = 0.01$ the difference is considered significant and we can be fairly confident in stating that there is a difference between the samples; from $p = 0.01$ to $p = 0.001$ the difference is highly significant and we can be very confident that there is a difference between the samples; if p is less than 0.001 (< 0.001) then the result is very highly significant and we can be almost certain that there is a difference between the samples.

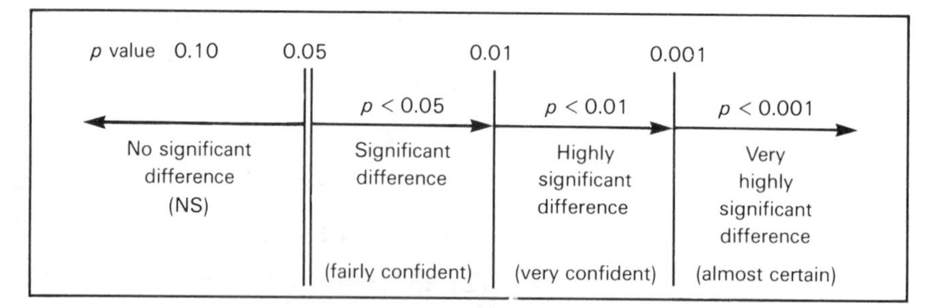

The graph of the t distribution (Appendix 3B) shows these significance levels clearly.

Having determined the value of p we can state what percentage of times we could expect to get such a result purely by chance: if $p = 0.001$ then once every thousand times (which is very unlikely); if $p = 0.05$ then 5 times in every 100, etc.

When we have analysed the results of an investigation we must relate back to the original parameters of the samples, particularly the means, and make a statement regarding the differences between them. It is not enough simply to say that there is a difference between the samples – the way in which they are different must also be explained. Examples of the whole process should make this clearer.

Example

Ten leaves were picked at random from a beech tree – five from the top branches and five from the lowest branches. It was hypothesised that the leaves from the highest branches would be larger than the lower ones since they would be better adapted to trapping the light where it is brightest.

Beech leaves

The lengths of the leaves in the two samples were measured carefully in centimetres and the mean and standard deviation of each sample were computed. The results were as below.

	Position on tree	
	Top (A)	**Bottom (B)**
Mean	6.2	7.6
Standard deviation	0.8	1.4

(i) **Using the null hypothesis**

1. State null hypothesis: there is no difference in the lengths of beech leaves from the top branches and the bottom branches.
2. Compute $t = \dfrac{6.2 - 7.6}{\sqrt{\dfrac{0.64}{5} + \dfrac{1.96}{5}}} = \dfrac{1.4}{0.72} = 1.94$
3. Determine the number of degrees of freedom $= (n_A - 1) + (n_B - 1) = 4 + 4 = 8$.
4. Using the table or graph of t distribution determine the value of p. p is greater than 0.05 (> 0.05).
5. Accept or reject the null hypothesis. We *accept* the null hypothesis since the p value is to the left of the 0.05 line.
6. There is no difference between the lengths of the two leaf samples – any differences are due to chance.

(ii) **Not using the null hypothesis**

1. Compute $t = 1.94$.
2. Determine the number of degrees of freedom = 8.
3. Determine the value of p from the table or graph > 0.05.
4. The difference between the lengths of the leaves is not significant (NS).

This was, however, a badly designed investigation since only five leaves were selected for each sample. The t test really requires sample sizes to be 30 or more. What would have happened if the two samples had contained 50 leaves each instead of five? Let us use the same values of the mean and standard deviation.

(i) **Using the null hypothesis**

1. $t = \dfrac{1.4}{0.228} = 6.14$

2. df = 49 + 49 = 98

3. $p < 0.001$

4. This time we *reject* the null hypothesis (there is no difference between the two samples), and the result is very highly significant.

5. There is therefore a difference between the two samples.

6. Looking at the mean values (6.2 for the leaves from the top of the tree and 7.6 for the leaves from the bottom of the tree) we can state that the lower leaves are *larger* than the upper leaves.

7. We must therefore still reject the original hypothesis which stated that the upper leaves would be larger than the lower ones. This conclusion can lead us to propose an alternative hypothesis which we can then test.

(ii) **Not using the null hypothesis**

1. $t = 6.14$
2. df = 98
3. $p < 0.001$
4. The difference between the samples is very highly significant.
5. The lower leaves are larger than the upper leaves.

EXERCISE

A new fertiliser for potatoes was advertised claiming higher yields. To test this claim, potatoes of the same variety were grown in 36 plots of a fixed size, 18 of the plots being treated with the usual fertiliser (A) and 18 plots with the new fertiliser (B).

The results are given as the yield in kilograms.

Fertiliser A	**27**	**20**	**16**	**18**	**22**	**19**	**23**	**21**	**17**	**28**	**19**	**17**	**26**	**22**	**19**	**23**	**22**	**20**
Fertiliser B	**28**	**19**	**19**	**21**	**24**	**20**	**25**	**19**	**27**	**25**	**29**	**23**	**21**	**17**	**23**	**27**	**25**	**24**

1 A cursory glance at these data would seem to indicate that the new fertiliser does produce higher yields. Using both methods of *t* test analysis (using the null hypothesis and not using it) determine if the claim of the advertiser is justified.

On occasions we might want to compare more than two sets of continuous data. In such cases a particular test of significance called 'analysis of variance' can be used, but the t test can suffice.

Example

Subjects were grouped according to how long they had been playing a musical instrument.

Non-Musicians

1–2 years
3–4 years
5–6 years
7–8 years
} Musicians

The subjects were tested for their ability to recall a note. The investigators wanted to find out if there was any significant difference in ability between the non-musicians and the musicians, and between the musicians who had been playing a musical instrument for differing periods of time.

In such cases a matrix is drawn to summarise the results.

		Musicians (time playing in years)			
		1–2	3–4	5–6	7–8
Non-Musicians		NS	< 0.05	< 0.01	< 0.001
Musicians	1–2		NS	NS	< 0.001
	3–4			NS	< 0.001
	5–6				< 0.01

There were ten t tests required to compare each group with all the other groups. The p values are given in each cell of the matrix so that you can see at a glance which tests were significant and which were not.

EXERCISE

Four varieties of rye-grass (A, B, C and D) were grown under identical conditions and after six weeks their heights were measured (using proper sampling techniques).

1 Construct a matrix to summarise the results of this experiment.

2 How many t tests would have to be carried out if each variety was to be compared with the other varieties?

A longitudinal section of the root of a broad bean was mounted on a slide and examined under the microscope. It was observed that the cells seemed to get larger the further they were from the tip.

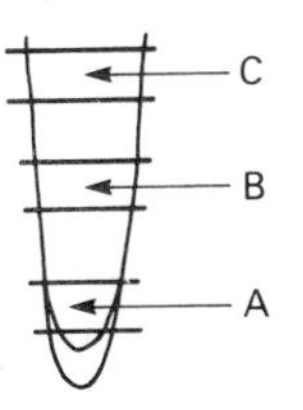

Longitudinal section of a broad bean root

In order to determine whether this was actually the case, ten cells were chosen at random from each of three areas – A, B and C. The length and width of each cell was measured using an eyepiece graticule.

The results were as below (in graticule divisions).

	Area A		Area B		Area C	
Cell no.	Length	Width	Length	Width	Length	Width
1	6.5	9.0	18.0	8.5	19.5	14.0
2	7.0	14.0	14.0	8.5	19.0	10.0
3	9.5	9.0	9.5	7.0	21.0	9.5
4	12.0	8.5	11.0	6.0	15.5	13.5
5	7.0	10.0	8.0	6.0	19.5	9.5
6	5.5	8.0	14.5	6.5	14.5	10.0
7	8.0	8.5	8.5	8.0	11.0	11.5
8	12.0	7.5	11.5	6.5	21.5	10.0
9	11.5	9.5	10.0	10.0	14.0	6.5
10	7.5	7.0	8.5	9.0	15.5	7.0

3 Carry out *t* tests to find out if there is a difference between the lengths of the cells in areas A, B and C.

4 Repeat this for the widths of the cells.

5 Suggest reasons for the results you obtain.

The χ^2 test

χ is the Greek letter chi, so χ^2 is pronounced ky-squared. This significance test was devised by the Englishman Karl Pearson in 1900. It is normally used when data consist of discrete variables (see p. 2). As with the *t* test we have to assume that the distribution of the expected frequencies approximates to a normal distribution. Again this means that if the expected frequencies are very small the validity of the test becomes questionable. The convention is that ***no expected frequency should be less than five.*** In certain cases results can be pooled to satisfy this condition.

The formula for χ^2 is quite simple. It is the sum of the observed frequency minus the expected frequency squared, divided by the expected frequency, abbreviated:

$$\chi^2 = \Sigma\frac{(O - E)^2}{E}$$

where Σ is the sum of; O is the observed frequency; E is the expected frequency.

There are a variety of uses of this test and examples of different situations that you are likely to come across are probably the best way to describe them.

Two categories (1 × 2 arrangement)

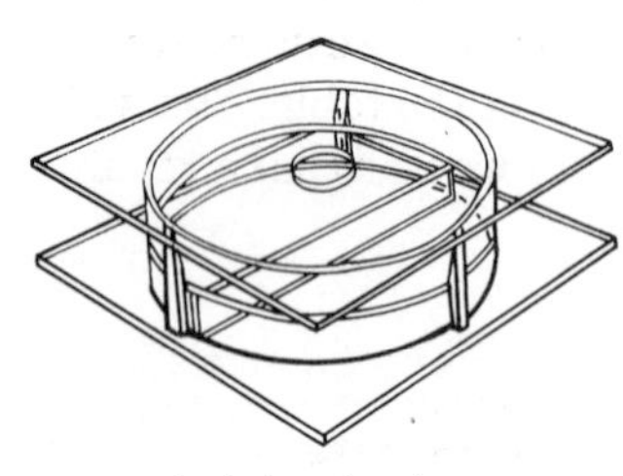
A choice chamber

Example

In a study of behaviour in the woodlouse a choice chamber was divided into two halves. One half was covered with black paper and the other half illuminated with a bench lamp. Ten woodlice were introduced through the opening and the stopper was replaced. After a few minutes when the woodlice had stopped moving the number in each half of the choice chamber were counted. The woodlice were then removed and the experiment was repeated five times with different woodlice.

The results were as in the table.

Experiment	Number of woodlice	
	In the dark	In the light
1	6	4
2	7	3
3	5	5
4	6	4
5	6	4

It would seem that the woodlice preferred the dark half of the choice chamber, but do these results indicate that there is a significant preference?

Totalling the woodlice in each half we find that 30 were found in the dark half and 20 in the light half. These are the observed values (O). We now compare these observed values with those that we would expect if the woodlice had not preferred any particular half, i.e. the number of woodlice would be the same in each half. This would be $50/2 = 25$: this is the expected number (E). The E values in such cases are determined on the basis of the null hypothesis, which in this case would be that there is no preference for either half of the choice chamber.

In situations like this where there are only two categories (a 1×2 arrangement) the accuracy of the χ^2 value is enhanced by using ***Yates' correction factor***, where 0.5 is subtracted from the absolute value (forget about minus signs) of the $(O - E)$ quantity before it is squared. Remember that it doesn't matter if $(O - E)$ is positive or negative, i.e. $-7.5 - 0.5 = -7.0$ and not -8.0.

The procedure is as follows.

1. Compute the value of χ^2. Always lay out the work in tabular form where possible since it will be neater and less liable to errors.

Category	O	E	$(O-E)$	$(O-E)-0.5$	$[(O-E)-0.5]^2$	$[(O-E)-0.5]^2/E$
Dark half	30	25	5	4.5	20.25	0.81
Light half	20	25	5	4.5	20.25	0.81
Totals	50	50				$\Sigma = 1.62$

$$\chi^2 = 1.62$$

2. Determine the number of degrees of freedom. This is equal to the number of categories minus 1. There were two categories (the dark half and the light half) so $2 - 1 = 1$ df.

3. Determine the value of p. To do this we follow the same procedure that was used for the t test but using the table or graph of the χ^2 distribution (see Appendix 4A and B). You will notice that this table differs from the t table in that more p values are given on the left of the table. You will see the reason for this later. Entering the table at df 1 we find that a χ^2 value of 1.62 lies between $p = 0.50$ and $p = 0.10$.

	Increase ←			*p* values		→ Decrease		
df	**0.99**	**0.95**	**0.90**	**0.50**	**0.10**	**0.05**	**0.01**	**0.001**
1	**0.000 16**	**0.0039**	**0.016**	**0.46**	**2.71**	**3.84**	**6.63**	**10.83**

1.62 lies here (between 0.46 and 2.71)

Not significant ← | → Significant (boundary at 3.84)

p is therefore greater than 0.10 (> 0.10).

4. Is this result significant or not? As in the t test the conventional level of significance is 5 per cent or $p = 0.05$ A p value of > 0.10 is therefore not significant. There was no preference for any particular side.

EXERCISE

1 The experiment was repeated but this time a lower wattage lamp was used. It was found that of the 50 woodlice, 33 preferred the dark side and 17 the light side. Carry out the χ^2 test to find out if these results are significant.

2 In another experiment a humidity gradient was established in the choice chamber by using water in one half and a drying agent in the other half of the bottom ring. Again 10 woodlice were placed in the choice chamber and the experiment was repeated five times. The results were as in the table on the next page.

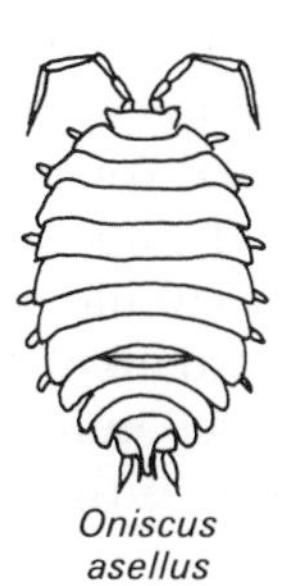
Oniscus asellus

	Experiment number				
	1	2	3	4	5
Damp half	5	6	8	7	5
Dry half	5	4	2	3	5

Find out if these results are significant. Do the woodlice prefer one particular half and, if so, which half?

More than two categories (1 × 3; 1 × 4; 1 × 5; etc)

Example

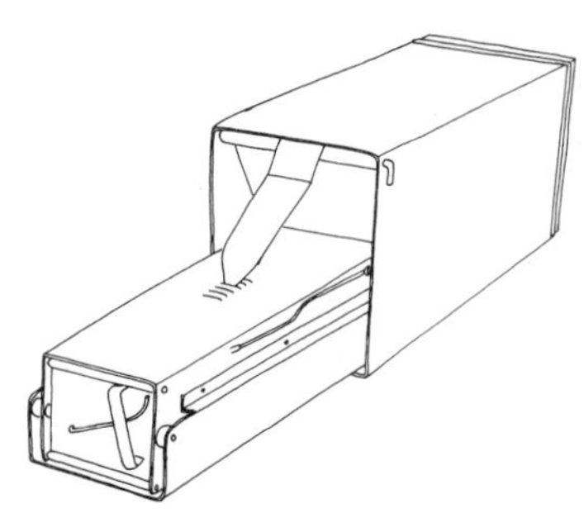
Longworth trap

Five Longworth (live) mammal traps were placed in randomly selected positions in a deciduous wood. The numbers of field mice captured in each trap over a period of time were recorded. After being caught the mice were released unharmed. In order to prevent the counting of the same mice more than once they were marked before release so that they could be recognised if recaptured.

The results were as in the table.

	Trap					
	A	B	C	D	E	Total
Number of field mice captured	22	26	21	8	23	100

From the data you should notice that trap D caught far fewer mice than any of the others. Did this happen by chance or is the result significant? If it is significant then we might be able to suggest hypotheses to explain such a result.

A total of 100 mice were caught. If there had been no difference in the number of mice caught in each of the traps (the null hypothesis) the expected numbers for each trap would be 100/5 = 20.

The information is tabulated in the usual way. Yates' correction factor is not used when there is more than one degree of freedom.

Trap	O	E	$(O - E)$	$(O - E)^2$	$(O - E)^2/E$
A	22	20	2	4	0.20
B	26	20	6	36	1.80
C	21	20	1	1	0.05
D	8	20	12	144	7.20
E	23	20	3	9	0.45
Totals	100	100			9.7

$$\chi^2 = 9.7$$

df = number of categories (traps) minus 1 = 5 − 1 = 4

The table or graph of χ^2 shows that 9.7 lies between $p = 0.05$ and 0.01. $p < 0.05$, so the result is significant. We can therefore be fairly sure that something other than chance has affected the number of mice caught in trap D.

EXERCISE

1 Try the test again, but this time omit trap D.

2 Try to think up as many explanations as possible for so few mice being captured by trap D.

Goodness-of-fit test

Sometimes we want to find out if the observed numbers agree with a hypothesised ratio or percentage. The expected numbers are calculated according to the hypothesised ratio and the χ^2 test can then be used to compare the observed numbers with the expected ones. This test is known as the 'goodness-of-fit' test and is often used in genetics. Genetic ratios are hypothetical, being based on Mendelian principles – actual observed numbers will depart from these ratios to a greater or lesser extent.

Two categories

Gregor Johann Mendel (1822–84)

Example

In one of his experiments Mendel raised 8023 plants that had arisen from the self-fertilisation of pea plants that were heterozygous for seed colour. Of these, 6022 were found to have yellow seeds and 2001 to have green seeds. Are these obtained numbers consistent with his hypothesised ratio of three dominant (yellow seeds) to one recessive (green seeds)?

We cannot reduce the observed numbers to a ratio and test this against the 3:1 ratio since any expected value must be greater than 5. We have to calculate the expected numbers according to the ratio. Since there are only two categories (yellow and green) and thus one degree of freedom, we should use Yates' correction factor, subtracting 0.5 from $(O - E)$ before squaring.

The total number of plants raised was 8023. A ratio of three parts to one part means four parts altogether. We divide the total number of plants by four: $8023/4 = 2005.75$. Three parts will therefore be $2005.75 \times 3 = 6017.25$, and one part will be 2005.75. (Note that decimals are used even though the test is dealing with whole numbers; rounding up can cause inaccuracies unless the values are very close to whole numbers.)

Category	O	E	$(O - E)$	$(O - E) - 0.5$	$[(O - E) - 0.5]^2$	$[(O - E) - 0.5]^2/E$
Yellow seeds	6022	6017.25	4.75	4.25	18.06	0.003
Green seeds	2001	2005.75	4.75	4.25	18.06	0.009
Totals	8023	8023.00				0.012

$\chi^2 = 0.012$

This is a very small value of χ^2 and we would expect such a difference not to be significant. The table (or graph) of χ^2, for 1 degree of freedom shows that the value of p is greater than 0.90 (> 0.90).

p	0.99	0.95	0.90	0.50
1 d.f.	0.000 16	0.0039	0.016	0.46

0.012 lies here (between 0.0039 and 0.016)

Since a p value greater than 0.05 means that any differences between the observed and expected values are not significant and are due to chance, we can say that the numbers that Mendel obtained agree with his hypothesised 3:1 ratio.

A word of caution is required. When the value of χ^2 is very small, corresponding to a high p value, i.e. when $p > 0.90$, the results should normally be re-examined. The observation fits the expectation so closely that one should suspect that a mistake has been made unconsciously or the experiment has been 'biased' in some way. A number of Mendel's results are like this and it has been suggested that one of his assistants could have been biased in his counting!

EXERCISE

1 In another monohybrid experiment Mendel found in the F2 generation that of a total of 1181 plants, 882 had inflated pods and 299 had wrinkled pods. Do these figures agree with his 3:1 ratio? Is there any indication of bias in this case?

Three categories

Example

A breeder used a particular strain of small mammal which had grey fur. Another breeder possessed a different strain which also had grey fur. When grey individuals of both strains were mated the offspring were found to be in a ratio of 1 black : 2 grey : 1 white. The white animals were weak and difficult to rear but if bred successfully together they produced offspring in the ratio of 1 black : 4 grey : 4 white. The breeders felt that these facts could be explained on the assumption that this is a case of dihybrid Mendelian inheritance but that a lethal influence is at work through some combination of recessive genes. In actual breeding crosses between a number of different white animals the total numbers were 12 black, 50 grey and 28 white. Do these figures agree with the hypothesised 1:4:4 ratio?

EXERCISE

1 Try to work out how the 1:4:4 ratio is arrived at – black is AABB and white AaBb.

The total number of animals was 12 + 50 + 28 = 90. The ratio of 1:4:4 has 9 parts so each ratio value is multiplied by 90/9, i.e. 10, to obtain the numbers expected in each category.

Category	O	E	$(O - E)$	$(O - E)^2$	$(O - E)^2/E$
Black	12	10	2	4	0.4
Grey	50	40	10	100	2.5
White	28	40	12	144	3.6
Totals	90	90			6.5

Since there are more than two categories, i.e. there is more than one degree of freedom, Yates' correction factor is not applied.

$\chi^2 = 6.5$

p	0.05	0.01
2 d.f.	5.99	9.21

6.5 lies here

df = number of categories minus one
= 3 (black, grey and white) − 1 = 2

In the χ^2 table at df 2, the value of 6.5 lies between $p = 0.05$ and $p = 0.01$, i.e. $p < 0.05$.

Since the value of p is less than 0.05 the result is significant – we can be fairly confident that the figures obtained deviate from a 1:4:4 ratio. The differences are due to something other than chance since we would only expect to get such a result less than 5 times in a 100. We can see from the numbers that there are fewer white ones than expected (28 as against 40). A clue is given at the start of the example: 'the white animals are weak and difficult to rear'. The shortfall of white mammals is possibly due to deaths. Taking this into account the figures could still agree with the 1:4:4 ratio

EXERCISE

In the common house mouse a number of genes have been found to interact, producing what appears to be a simple character. When black-coated mice were crossed with white-coated mice the resultant 64 offspring consisted of 11 black, 14 white and 39 agouti coloured mice. Agouti is a sort of grey colour that is characteristic of wild mice.

1 The expected ratio according to Mendelian theory is 9 agouti : 3 black : 4 white. Do the numbers obtained agree with this ratio?

Four categories

The same procedure is followed as in the case of three categories.

Example

A farmer who was involved in egg production liked to keep in touch with the latest developments on the subject so he borrowed a recently published book on poultry farming from the library. He found in the book that his hens should be producing eggs in the following percentages.

> Large 40 per cent; standard 45 per cent; medium 10 per cent; small 5 per cent

He decided to check how his own hens compared with these percentages. His egg production for the following week was as follows.

> Large 590; standard 652; medium 112; small 46

How do the farmer's hens compare with the figures given in the book? In this case we can carry out a goodness-of-fit test to find out.

First we have to work out the number of eggs that would be expected according to the percentages given in the book.

The total number of eggs produced = 590 + 652 + 112 + 46 = 1400

$$
\begin{aligned}
40 \text{ per cent} &= 1400 \times 40/100 = 560 \\
45 \text{ per cent} &= 1400 \times 45/100 = 630 \\
10 \text{ per cent} &= 1400 \times 10/100 = 140 \\
5 \text{ per cent} &= 1400 \times 5/100 = 70 \\
\text{Total} &= 1400 \text{ (as a check on the calculation)}
\end{aligned}
$$

The information is tabulated.

Categories (type of egg)	Observed	Expected
Large	**590**	**560**
Standard	**652**	**630**
Medium	**112**	**140**
Small	**46**	**70**
Totals	1400	1400

From a cursory glance at this table it would seem that the farmer's hens are doing better than those described in the book, but are the differences significant? Remember also that the farmer only counted his egg production for 1 week – it could be that it was a particularly good week.

The χ^2 test is carried out.

Type of egg	O	E	$(O - E)$	$(O - E)^2$	$(O - E)^2/E$
Large	**590**	**560**	**30**	**900**	**1.60**
Standard	**652**	**630**	**22**	**484**	**0.77**
Medium	**112**	**140**	**28**	**784**	**5.60**
Small	**46**	**70**	**24**	**576**	**8.23**
Totals	1400	1400			16.20

$$\chi^2 = 16.20$$
$$df = 4 - 1 = 3$$
$$p < 0.01$$

This value of p is of course highly significant, and we can be very confident that there is a difference between the farmer's egg production and that given in the book. Referring back to the observed and expected numbers we can see that the farmer had more large and standard eggs and fewer medium and small. The farmer would be advised to continue his own methods but he should check his egg production more often.

EXERCISE

In one of Mendel's dihybrid crosses, the following types and numbers of pea plants were recorded in the F2 generation.

Yellow and round seeds	**315**
Yellow and wrinkled seeds	**101**
Green and round seeds	**108**
Green and wrinkled seeds	**32**

1 According to theory these should be in the ratio of 9:3:3:1. Do these obtained numbers agree with the ratio?

Homogeneity test

We saw that some of Mendel's results were so close to his anticipated ratios that we suspect something untoward could have been occurring. Unconscious bias can influence judgement. Small values of χ^2 that give p values greater than 0.90 indicate the possibility of such a kind of bias. On the other hand large values of χ^2 can indicate irregularities in technique.

The χ^2 test is therefore useful in deciding if individual results in a class can be grouped; if replicates (repetitions of an experiment) are similar enough to be used collectively; if a particular technique has been adequately mastered by different groups of students, etc.

Example

In a class experiment students studied the growth of a population of the yeast *Schizosaccharomyces pombe*. At given intervals of time, each student had to take a homogenised sample from the culture vessel using a special pipette and add it to a haemocytometer slide following precise instructions. These instructions included one which exhorted students to be particular about the technique so that consistent results would be obtained. The students had to count the number of yeast cells in each of five squares on the slide under the microscope using a precise counting technique.

If the culture had been theoretically homogeneous and the techniques absolutely flawless one would have expected no difference between the numbers of yeast cells in each of the squares at any particular time.

Student A's results for the first time interval were as follows.

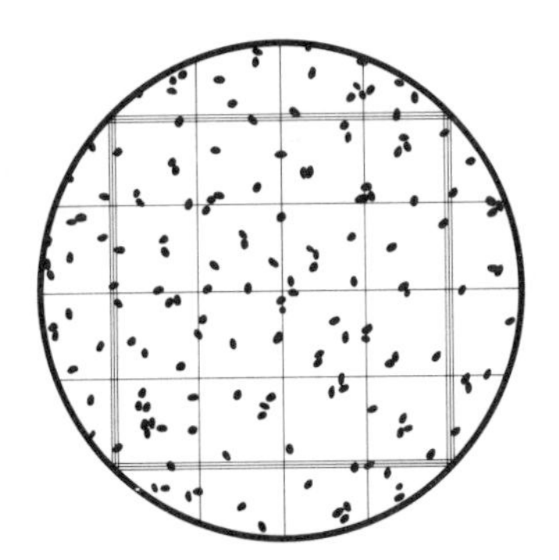

Yeast cells on a haemocytometer grid (× 40 objective)

Square number	Number of yeast cells
1	65
2	78
3	56
4	71
5	60

Total 330

If there had been no difference between the counts for the squares the number of cells in each square would be 330/5 = 66 yeast cells.

We now carry out the χ^2 test in the usual way.

Square number	O	E	$(O - E)$	$(O - E)^2$	$(O - E)^2/E$
1	65	66	1	1	0.015
2	78	66	12	144	2.180
3	56	66	10	100	1.520
4	71	66	5	25	0.380
5	60	66	6	36	0.550
Totals	330	330			4.65

$$\chi^2 = 4.65$$

$$\text{df} = 5 - 1 = 4$$

$p > 0.10$ since it lies between 0.50 (3.36) and 0.10 (7.78), so there is no significant difference between the results, any differences being due to chance. This particular student's technique seems to be adequate.

Student B's results for the same time interval were as follows.

Square number	Number of yeast cells
1	42
2	75
3	89
4	53
5	31

Total 290

The expected value if there is no difference between the squares = 290/5 = 58.

Student B's results seem more variable than A's but are the differences between the squares significant or not?

Square number	O	E	$(O - E)$	$(O - E)^2$	$(O - E)^2/E$
1	42	58	16	256	4.41
2	75	58	17	289	4.98
3	89	58	31	961	16.57
4	53	58	5	25	0.43
5	31	58	27	729	12.57
Totals	290	290			38.96

$$\chi^2 = 38.96$$

For df 4, this value of χ^2 gives a p value of much less than 0.001. There is a very highly significant difference between the squares – so much so that we could have serious doubts about the technique being used. Such a set of counts should not be included when combining different students' results.

Student C's results for the same time interval were different again.

Square number	Number of yeast cells
1	67
2	68
3	67
4	66
5	67
	Total 335

These results appear to be somewhat suspicious. The χ^2 test will indicate if they are too good to be true.

Square number	O	E	$(O-E)$	$(O-E)^2$	$(O-E)^2/E$
1	67	67	0	0	0
2	68	67	1	1	0.015
3	67	67	0	0	0
4	66	67	1	1	0.015
5	67	67	0	0	0
Totals	335	335			0.03

$$\chi^2 = 0.03$$

This χ^2 value is very small, and for df 4 gives a p value much greater than 0.99. Student C obviously over-reacted to the instruction that the results should be consistent. Such a set of values could never be obtained even if all the instructions had been carefully followed. Again, since these readings are suspect, they should not be included in the class data.

EXERCISE

In the experiment on yeast cell population growth, the mean numbers of yeast cells at the first time interval for 10 students were as follows.

Student	1	2	3	4	5	6	7	8	9	10
Mean number of yeast cells (rounded up)	68	76	64	48	78	82	71	65	110	72

1 Are these results homogeneous enough to be grouped together? Use the test of homogeneity to find out.

Contingency tables

In the examples and problems of χ^2 so far considered there are several categories on attributes into which the observations can be placed. These are then tabulated. Such tables are described by the number of columns and rows they contain – so far they have been of the one column type; $1 \times n$ (1 column and n rows) – 1×2, 1×3, 1×4, 1×5, etc.

On p. 11 you saw that if there are two sets of observations and two categories, so that there are four possible combinations, the data can be set out as a 2×2 contingency table.

Example

Cepaea nemoralis, the banded snail, was collected from two localities, a beechwood and from under hedges. Of 150 snails collected from the beechwood 40 were found to be banded and the other 110 unbanded. Of 70 snails from under the hedges, 45 were banded and 25 unbanded. What we want to know is whether there is any relationship between banding and the location.

The data are entered in the table as below:

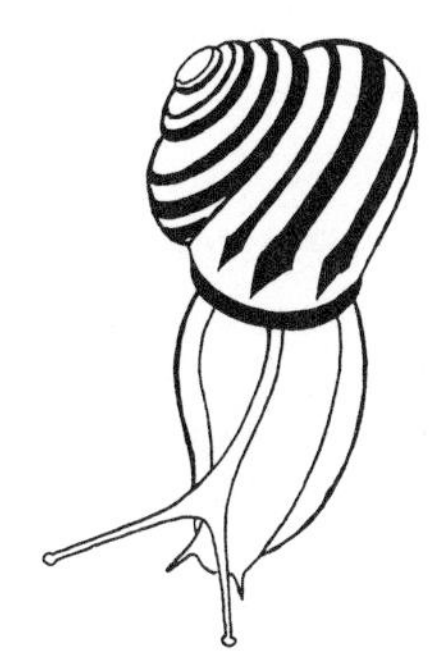

Capaea nemoralis

Location	Type of snail		Totals
	Banded	Unbanded	
Beechwood	40	110	150
Under hedges	45	25	70
Totals	85	135	220 (Grand total)

To calculate the value of χ^2 we have to determine the expected values that correspond to the four observed values, based on the assumption that there is no relationship between the type of snail and the location (null hypothesis). For each cell of the table we multiply the column total by the row total and divide by the grand total e.g. for the top left-hand cell (banded snails from the beechwood – let us call it cell a):

Location	Type of snail		Row totals
	Banded	Unbanded	
Beechwood	40		150
Under hedges			
Column totals	85		220 (Grand total)

$$\text{Expected value} = \frac{\text{column total} \times \text{row total}}{\text{grand total}} = \frac{85 \times 150}{220} = 57.95$$

Since this figure is so near to 58 we shall round it up for simplicity.

Following the rule for determining the *E* values we can calculate them for the other cells (b, c and d).

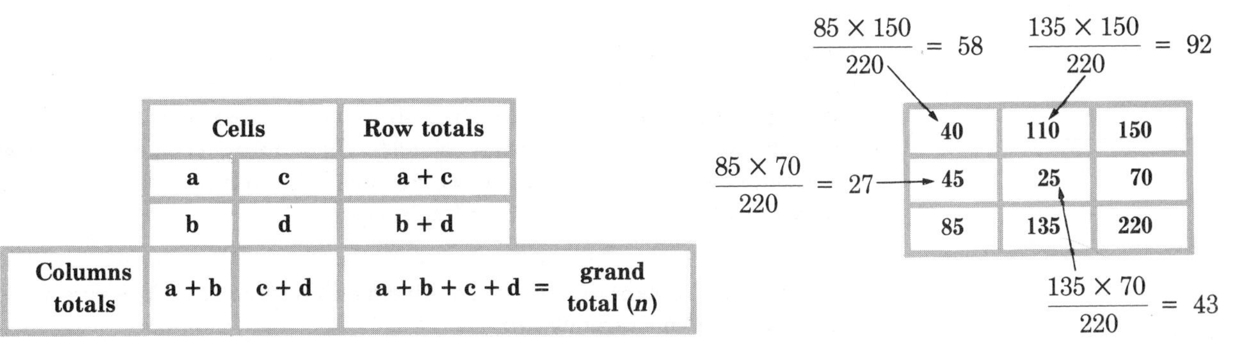

	Cells		Row totals
	a	c	a + c
	b	d	b + d
Columns totals	a + b	c + d	a + b + c + d = grand total (n)

Actually when the first cell value has been determined the others can be obtained by subtraction since the totals are fixed.

58	150 − 58 = 92	150
85 − 58 = 27	70 − 27 or 135 − 92 = 43	70
85	135	

Only one of the cells is thus free to vary. All the others depend on its value and so there is only one degree of freedom. The formula for the degrees of freedom is therefore the number of columns minus one multiplied by the number of rows minus one $(c - 1) \times (r - 1)$. In a 2×2 contingency table there are two columns and two rows so the number of degrees of freedom will be $(2 - 1) \times (2 - 1) = 1 \times 1 = 1$ df.

We can now enter both the observed and expected values in the table. The *E* values can be conveniently placed in brackets.

Location	Type of snail		Totals
	Banded	Unbanded	
Beechwood	40 (58)	110 (92)	150
Under hedges	45 (27)	25 (43)	70
Totals	85	135	220

Since there is only one degree of freedom, Yates' correction factor should be used, i.e. subtracting 0.5 from $(O - E)$ for each cell before squaring.

Cell	O	E	$(O - E)$	$(O - E) - 0.5$	$[(O - E) - 0.5]^2$	$[(O - E) - 0.5]^2/E$
A	40	58	18	17.5	306.25	5.28
B	45	27	18	17.5	306.25	11.34
C	110	92	18	17.5	306.25	3.33
D	25	43	18	17.5	306.25	7.12
Totals	220	220				27.07

$$\chi^2 = 27.07$$

$$\text{df} = (2 - 1) \times (2 - 1) = 1$$

The value of p when χ^2 is 27.07 and df = 1 is < 0.001, so the result is very highly significant. We can be almost certain that there is a difference in the numbers of banded and unbanded snails from the two locations.

From examination of the contingency table above we can see that there were more banded snails from under the hedges than we would have expected and fewer from the beechwood; and more unbanded ones from the beechwood and less from under the hedges. Such a finding can lead us to suggest hypotheses to explain the result. Have you any ideas?

There is an alternative method of computing the χ^2 value for 2×2 contingency tables (and only 2×2 tables!).

The formula is

$$\chi^2 = \frac{n(|ad - bc| - \frac{1}{2}n)^2}{(a + b)\,(c + d)\,(a + c)\,(b + d)}$$

where n is the grand total, and $|ad - bc|$ represents an absolute number, i.e. forget about any minus sign.

Unfortunately this method generates very large numbers which most calculators cannot cope with. It is therefore only suitable when each cell contains very small numbers. Try the banded snail example above using this formula.

EXERCISE

The thrush is a common predator of *Cepaea*. It smashes the shells open by holding them in its beak and hitting them against a stone, which is known as the anvil. Shells of *Cepaea* were collected and grouped according to whether they fell into each of the following categories:

> broken and banded, broken and unbanded,
> intact and banded, intact and unbanded.

The numbers were as in the table.

	Banded	Unbanded
Broken	68	30
Intact	342	195

1 Use the χ^2 test to find out if there is any relationship between banding and predation by the thrush.

2 Determine the value of χ^2 using the alternative formula for 2 × 2 tables.

Thrush breaking banded snails on an 'anvil' stone

The test of association

The 2 × 2 contingency table can be used to find out whether there is any association between two species in a community or habitat – does the presence of one species make it more likely that the other species will also be present? Perhaps the soil conditions suit both species, or one of the species might provide favourable conditions for the other. The presence of one species might prevent the presence of the other species – this would be an example of a negative association. In many cases the association might not be immediately obvious so that it would be necessary to sample the habitat and analyse the results using χ^2.

We can sample the habitat by placing quadrats at selected random positions and noting whether species A and B are absent or present. There are therefore four possibilities:

> species A – present or absent;
> species B – present or absent.

Three different situations can arise – no association; a positive association, where the presence of one species makes it more likely that the other species will also occur; a negative association, where the presence of one species makes it less likely than the other species will also occur. Evidence for such associations, which may or may not be obvious from direct observations, can lead to hypotheses as to their cause.

(i) **No association**
Where there are equal (or nearly equal) numbers in the columns

Species B	Species A		
	Absent	Present	Totals
Present	51	74	125
Absent	49	76	125
Totals	100	150	250

(ii) **Positive association**
Where there are greater numbers in the present/present cell and in the absent/absent cell

Species B	Species A		
	Absent	Present	Totals
Present	10	115	125
Absent	90	35	125
Totals	100	150	250

(iii) **Negative association**
Where there are greater numbers in the present/absent cells

Species B	Species A		
	Absent	Present	Totals
Present	80	45	125
Absent	20	105	125
Totals	100	150	250

χ^2 is calculated in the usual way for a 2×2 contingency table.

EXERCISE
One hundred quadrat frames were placed at random on a lawn and scored as to whether daisies and dandelions were present or absent. The results are presented as a 2×2 contingency table.

Dandelions	Daisies	
	Absent	Present
Present	10	40
Absent	35	15

1 Is there an association between the daisies and the dandelions on this lawn?

2 If there is an association, what type of association is it?

Combined 2 × 2 contingency table and test of homogeneity

Occasionally one might want to compare two particular sets of observations in order to find out if they are significantly different or not.

Example

A class was divided into two groups (A and B) to carry out a breeding experiment using the fruit-fly *Drosophila*. Pure-breeding normal-winged flies were crossed with pure-breeding vestigial-winged flies and these produced an F1 generation consisting of normal-winged flies. When these were interbred the numbers of phenotypes in the F2 generation were as below.

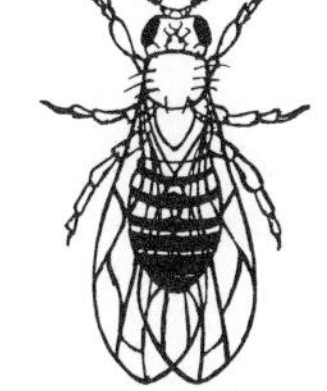

normal-winged male

Drosophila melanogaster

vestigial-winged female

Group A: normal-winged, 92; vestigial-winged, 29
Group B: normal-winged, 96; vestigial-winged, 14

Are these two sets of results significantly different from each other even though both groups followed the same instructions?

The data were laid out as a 2 × 2 contingency table.

	Normal-winged	Vestigial-winged	Totals
Group A	92 (98.5)	29 (22.5)	121
Group B	96 (89.5)	14 (20.5)	110
Totals	188	43	231

The expected values were computed in the usual way (see p. 105) and are shown in brackets.

Cell	O	E	$(O-E)$	$(O-E)-0.5$	$[(O-E)-0.5]^2$	$[(O-E)-0.5]^2/E$
A	92	98.5	6.5	6.0	36.0	0.37
B	96	89.5	6.5	6.0	36.0	0.40
C	29	22.5	6.5	6.0	36.0	1.60
D	14	20.5	6.5	6.0	36.0	1.76
Totals	231	231.0				4.13

$$\chi^2 = 4.13$$

$$\text{df} = 1, \text{ i.e. } (c-1) \times (r-1)$$

From the table of χ^2, p is less than 0.05 ($p < 0.05$) since $4.13 > 3.84$. There is therefore a significant difference between the numbers obtained by the two class groups.

According to Mendelian theory we would expect a ratio of 3 normal : 1 vestigial in the F2 generation. The ratio of normal to vestigial for group A works out at 3.17:1, whereas it is 6.86:1 for group B. From the contingency table it would appear that there are far fewer vestigial-winged flies than would be expected for group B. χ^2 tests for each of the groups' observed numbers would probably indicate no significant difference between those of group A and a 3:1 ratio, but those for group B could deviate significantly. You could carry out these two χ^2 tests to find out!

One suggestion or hypothesis to explain the short-fall of vestigial-winged flies might be that the vestigial-winged flies got caught in the food medium because they couldn't fly, and so they died.

Perhaps Group B forgot to keep the culture tube on its side while the flies recovered from the ether. This could be checked and the experiment repeated if necessary.

c × r contingency tables

There can be any number of columns and rows in a contingency table. The interpretation of the information in a 2 × 2 table was found to be reasonably straightforward (it is hoped) but in more complex tables interpretation can be more difficult.

Example

Let us look first of all at a 3 × 2 table. In an investigation of a certain disease the following data were obtained concerning the numbers of patients with different symptoms, according to the concentration of a particular substance in their blood. There are three columns representing no symptoms, mild symptoms and severe symptoms. The concentration in the blood is either high or low. We will use the test to find out if there is any connection between the severity of the symptoms and the concentration in the blood.

OBSERVED

Concentration	Symptoms			
in the blood	None	Mild	Severe	Totals
High	16	16	12	44
Low	30	10	6	46
Totals	46	26	18	90

The expected values are worked out in the same manner as in the 2 × 2 table.

EXPECTED

Concentration in the blood	Symptoms			Totals
	None	Mild	Severe	
High	$\frac{46 \times 44}{90} = 22.5$ *	$\frac{26 \times 44}{90} = 12.7$ *	$44 - (22.5 + 12.7) = 8.8$	44
Low	$46 - 22.5 = 23.5$	$26 - 12.7 = 13.3$	$18 - 8.8 = 9.2$	46
Totals	46	26	18	90

Only two cells (those marked with an *) were obtained by multiplication and division – the rest were obtained by subtraction from the totals, so there are two degrees of freedom. There were three columns and two rows, so $(3 - 1) \times (2 - 1) = 2 \times 1 = 2$ df.

Since there is more than one degree of freedom Yates' correction factor is not used.

O	E	$(O-E)$	$(O-E)^2$	$(O-E)^2/E$
16	22.5	6.5	42.25	1.88
30	23.5	6.5	42.25	1.80
16	12.7	3.3	10.89	0.86
10	13.3	3.3	10.89	0.82
12	8.8	3.2	10.24	1.16
6	9.2	3.2	10.24	1.11
Totals 90	90.0			7.63

$$\chi^2 = 7.63$$

For df 2, $p < 0.05$. This result is significant. There is a relationship between the symptoms and the concentration in the blood. Looking at the table we can see that the higher the concentration the more severe are the symptoms.

EXERCISE

In an investigation to find out if there is any relationship between eye colour and colour preference, 536 students were classified into five categories of eye colour thus:

blue; green; grey; hazel; brown;

and into four categories of colours preferred, i.e. each student selected the colour they most preferred from the four shown:

red; blue; green; yellow.

The results were as follows:

Eye colour	Colour preferred				Totals
	Red	Blue	Green	Yellow	
Blue	97	144	20	32	293
Green	16	38	7	11	72
Grey	17	22	5	7	51
Hazel	12	32	6	8	58
Brown	10	34	8	10	62
Totals	152	270	46	68	536

1 What is the χ^2 value for these data?
2 Is there any relationship between eye colour and colour preference?

The correlation coefficient test

In section 3.1.2, p. 59 you found out how to determine the correlation coefficient 'r', which gave a figure between +1 and −1 to indicate how close the correlation was between the variables. Since calculation of r assumes that both variables are normally distributed it is possible to determine the significance of the correlation. The significance levels depend, however, on the size of the sample.

We can determine the level of significance of any value of r in a similar manner to that for t and χ^2, using a table or graph. The table of the correlation coefficient is given in Appendix 5A and the graph in Appendix 5B.

The number of degrees of freedom is two less than the number of values of either variable $(n - 2)$. If $p < 0.05$ then the correlation is significant; if p lies between 0.05 and 0.01 the result is significant and we can be fairly confident that a correlation exists; if p lies between 0.01 and 0.001 then the result is highly significant and we can be very confident that there is a correlation between the variables; if $p < 0.001$ then of course the result is very highly significant and we can be almost certain that a correlation exists.

Examine the table of the correlation coefficient (Appendix 5A). You will find the usual layout with p levels along the top and degrees of freedom down the left-hand side. As you move down the table, i.e. as the sample size increases, the values of r get smaller. In fact you may be surprised to find that at large sample sizes the value of r can be relatively small and still indicate a significant correlation between the variables, e.g. when df $= 100$ an r value of 0.2 is significant. On the other hand when the samples are small r must be close to 1 to be significant. Significance is treated in the same way whether r is positive or negative – the conclusion is simply a significant positive or negative correlation.

Example
On p. 56 a problem was given concerning two variables. A scattergraph was then drawn from the data given on Work Sheet 45, and from this a value of r had to be estimated. The actual calculated value of r from the data is -0.77. Is this value of r significant or not?

There are 30 observations for both variables so there are $30 - 2 = 28$ degrees of freedom. Entering the table of the correlation coefficient (Appendix 5A) at df 28 (the nearest is 30) we find that the value of 0.77 lies to the right of $p = 0.001$, so $p < 0.001$, the result is very highly significant and we can be almost certain that a negative (since $r = -0.77$) correlation exists between the two variables.

EXERCISE

1 Determine whether the r value obtained in question 1, p. 60 (Work Sheet 17) is significant or not.

2 Determine whether the r value obtained in question 3, p. 60 (Work Sheet 46) is significant or not.

3 Using the privet data given on Work Sheet 47, compute the value of r and find out if there is a significant correlation between the length and breadth of privet leaves.

Appendixes

APPENDIX 1

Statistical symbols and formulae

Symbol	Definition	Description	Computational form
x		The value of any particular variate	
n	sample size	The number of observations in a sample	
f	frequency	The number of times a particular observation occurs	
m	the gradient	The steepness of a graph line	
C	the intercept	The value of Y when $X = 0$	
$<$	is less than		
$>$	is greater than		
Σ	capital sigma	The sum of any specified values	
df (or v)	degrees of freedom	The number of values, attributes, etc., that are free to vary	$n - 1$; $(c - 1) \times (r - 1)$ etc.
p	probability value	The value determined by a test of significance, where totality is 1	
$\bar{x}$ (x-bar)	sample mean or average	Total of observed values in the sample divided by the number of observations	$\frac{\Sigma x}{n}$ for ungrouped data; $\frac{\Sigma fx}{\Sigma f}$ for grouped data
$(x - \bar{x})$	deviation from the mean	The difference between any particular observation and the mean	$x - \bar{x}$
s^2	sample variance	The sum of the squares of the deviations from the mean divided by the number of observations minus one	$\frac{\Sigma(x - \bar{x})^2}{n - 1}$ for ungrouped data; $\frac{\Sigma f(x - \bar{x})^2}{\Sigma f - 1}$ for grouped data
s	sample standard deviation	The positive square root of the sample variance	$\sqrt{\frac{\Sigma(x - \bar{x})^2}{n - 1}}$; $\sqrt{\frac{\Sigma f(x - \bar{x})^2}{\Sigma f - 1}}$
f_{cum}	cumulative frequency	The addition of a succeeding value to a preceding one	f_1; $f_1 + f_2$; $f_1 + f_2 + f_3$; etc.

Symbol	Definition	Description	Computational form
Percentage f_{cum}	cumulative percentage	The addition of a succeeding percentage to a preceding one	$\%_1$; $\%_1 + \%_2$; $\%_1 + \%_2 + \%_3$; etc.
Z (or d)	standardised normal deviate	Variate minus the mean divided by the standard deviation	$\frac{x - \bar{x}}{s}$
$\hat{\sigma}^2_{\bar{x}}$	variance of the mean	The variance divided by the number of observations	$\frac{s^2}{n}$
$\hat{\sigma}_{\bar{x}}$ or S.E.	standard deviation of the mean (standard error)	The standard deviation divided by the square root of the number of observations The square root of the variance of the mean	$\frac{s}{\sqrt{n}}$ $\sqrt{\frac{s^2}{n}}$
	confidence limits	The mean plus or minus the value of t for a given number of degrees of freedom times the standard deviation of the mean	$\bar{x} \pm t\hat{\sigma}_{\bar{x}} = \bar{x} \pm t \times \frac{s}{\sqrt{n}}$ $= \bar{x} \pm t \times \sqrt{\frac{s^2}{n}}$
$\hat{\sigma}^2(\bar{x}_A - \bar{x}_B)$	variance of the difference between sample means	The variance of the mean of sample A plus the variance of the mean of sample B	$\hat{\sigma}^2_{\bar{x}_A} + \hat{\sigma}^2_{\bar{x}_B} = \frac{s^2_A}{n_A} + \frac{s^2_B}{n_B}$
$\hat{\sigma}(\bar{x}_A - \bar{x}_B)$	standard deviation of the difference between sample means	The standard deviation of the mean of sample A plus the standard deviation of the mean of sample B; the square root of the variance of the difference of sample means	$\hat{\sigma}_{\bar{x}_A} + \hat{\sigma}_{\bar{x}_B}$ $= \sqrt{\hat{\sigma}^2_{\bar{x}_A} + \hat{\sigma}^2_{\bar{x}_B}}$ $= \sqrt{\frac{s^2_A}{n_A} + \frac{s^2_B}{n_B}}$
t	Student's t significance test	The mean of sample A minus the mean of sample B divided by the standard deviation of the difference between sample means; the mean of sample A minus the mean of sample B divided by the square root of the variance of the difference between sample means	$\frac{\bar{x}_A - \bar{x}_B}{\hat{\sigma}_{\bar{x}_A} + \hat{\sigma}_{\bar{x}_B}} = \frac{\bar{x}_A - \bar{x}_B}{\sqrt{\hat{\sigma}^2_{\bar{x}_A} + \hat{\sigma}^2_{\bar{x}_B}}}$ $= \frac{\bar{x}_A - \bar{x}_B}{\sqrt{\frac{s^2_A}{n_A} + \frac{s^2_B}{n_B}}}$
χ^2	chi-squared significance test	The square of the difference between the observed and expected numbers divided by the expected number in each class; then the sum of the quotients	$\left[\frac{(O - E)^2}{E}\right]_1 +$ $\left[\frac{(O - E)^2}{E}\right]_2 + \ldots$ $\left[\frac{(O - E)^2}{E}\right]_n +$
r	product moment correlation coefficient	+1 is perfect positive correlation; 0 no correlation; −1 perfect negative correlation	$\frac{\Sigma(X - \bar{X})(Y - \bar{Y})}{\sqrt{\Sigma(X - \bar{X})^2 \Sigma(Y - \bar{Y})^2}}$

APPENDIX 2

Areas of the standardised normal distribution

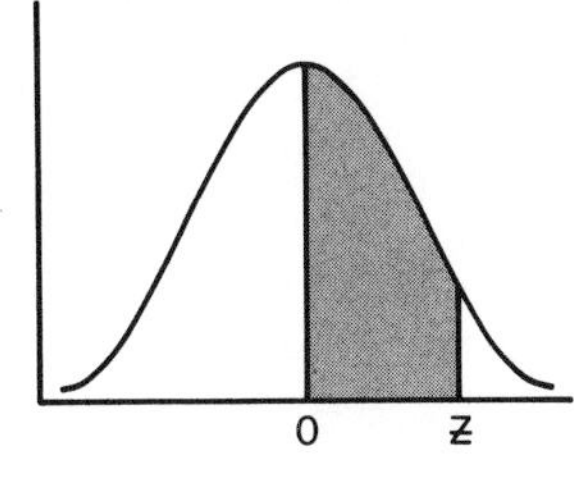

An entry in the table is the proportion under the entire curve which is between $Z = 0$ and a positive value of Z. Areas for negative values of Z are obtained by symmetry.

Z	.00	.01	.02	.03	.04	.05	.06	.07	.08	.09
0.0	0.0000	0.0040	0.0080	0.0120	0.0160	0.0199	0.0239	0.0279	0.0319	0.0359
0.1	0.0398	0.0438	0.0478	0.0517	0.0557	0.0596	0.0636	0.0675	0.0714	0.0753
0.2	0.0793	0.0832	0.0871	0.0910	0.0948	0.0987	0.1026	0.1064	0.1103	0.1141
0.3	0.1179	0.1217	0.1255	0.1293	0.1331	0.1368	0.1406	0.1443	0.1480	0.1517
0.4	0.1554	0.1591	0.1628	0.1664	0.1700	0.1736	0.1772	0.1808	0.1844	0.1879
0.5	0.1915	0.1950	0.1985	0.2019	0.2054	0.2088	0.2123	0.2157	0.2190	0.2224
0.6	0.2257	0.2291	0.2324	0.2357	0.2389	0.2422	0.2454	0.2486	0.2517	0.2549
0.7	0.2580	0.2611	0.2642	0.2673	0.2703	0.2734	0.2764	0.2794	0.2823	0.2852
0.8	0.2881	0.2910	0.2939	0.2967	0.2995	0.3023	0.3051	0.3078	0.3106	0.3133
0.9	0.3159	0.3186	0.3212	0.3238	0.3264	0.3289	0.3315	0.3340	0.3365	0.3389
1.0	0.3413	0.3438	0.3461	0.3485	0.3508	0.3531	0.3554	0.3577	0.3599	0.3621
1.1	0.3643	0.3665	0.3686	0.3708	0.3729	0.3749	0.3770	0.3790	0.3810	0.3830
1.2	0.3849	0.3869	0.3888	0.3907	0.3925	0.3944	0.3962	0.3980	0.3997	0.4015
1.3	0.4032	0.4049	0.4066	0.4082	0.4099	0.4115	0.4131	0.4147	0.4162	0.4177
1.4	0.4192	0.4207	0.4222	0.4236	0.4251	0.4265	0.4279	0.4292	0.4306	0.4319
1.5	0.4332	0.4345	0.4357	0.4370	0.4382	0.4394	0.4406	0.4418	0.4429	0.4441
1.6	0.4452	0.4463	0.4474	0.4484	0.4495	0.4505	0.4515	0.4525	0.4535	0.4545
1.7	0.4554	0.4564	0.4573	0.4582	0.4591	0.4599	0.4608	0.4616	0.4625	0.4633
1.8	0.4641	0.4649	0.4656	0.4664	0.4671	0.4678	0.4686	0.4693	0.4699	0.4706
1.9	0.4713	0.4719	0.4726	0.4732	0.4738	0.4744	0.4750	0.4756	0.4761	0.4767
2.0	0.4772	0.4778	0.4783	0.4788	0.4793	0.4798	0.4803	0.4808	0.4812	0.4817
2.1	0.4821	0.4826	0.4830	0.4834	0.4838	0.4842	0.4846	0.4850	0.4854	0.4857
2.2	0.4861	0.4864	0.4868	0.4871	0.4875	0.4878	0.4881	0.4884	0.4887	0.4890
2.3	0.4893	0.4896	0.4898	0.4901	0.4904	0.4906	0.4909	0.4911	0.4913	0.4916
2.4	0.4918	0.4920	0.4922	0.4925	0.4927	0.4929	0.4931	0.4932	0.4934	0.4936
2.5	0.4938	0.4940	0.4941	0.4943	0.4945	0.4946	0.4948	0.4949	0.4951	0.4952
2.6	0.4953	0.4955	0.4956	0.4957	0.4959	0.4960	0.4961	0.4962	0.4963	0.4964
2.7	0.4965	0.4966	0.4967	0.4968	0.4969	0.4970	0.4971	0.4972	0.4973	0.4974
2.8	0.4974	0.4975	0.4976	0.4977	0.4977	0.4978	0.4979	0.4979	0.4980	0.4981
2.9	0.4981	0.4982	0.4982	0.4983	0.4984	0.4984	0.4985	0.4985	0.4968	0.4986
3.0	0.4987	0.4987	0.4987	0.4988	0.4988	0.4989	0.4989	0.4989	0.4990	0.4990

APPENDIX 3A

Table of *t* distribution

	decreasing value of p ⟶			
Degrees of freedom (df)	p values			
	0.10	0.05	0.01	0.001
1	6.31	12.71	63.66	636.60
2	2.92	4.30	9.92	31.60
3	2.35	3.18	5.84	12.92
4	2.13	2.78	4.60	8.61
5	2.02	2.57	4.03	6.87
6	1.94	2.45	3.71	5.96
7	1.89	2.36	3.50	5.41
8	1.86	2.31	3.36	5.04
9	1.83	2.26	3.25	4.78
10	1.81	2.23	3.17	4.59
12	1.78	2.18	3.05	4.32
14	1.76	2.15	2.98	4.14
16	1.75	2.12	2.92	4.02
18	1.73	2.10	2.88	3.92
20	1.72	2.09	2.85	3.85
22	1.72	2.08	2.82	3.79
24	1.71	2.06	2.80	3.74
26	1.71	2.06	2.78	3.71
28	1.70	2.05	2.76	3.67
30	1.70	2.04	2.75	3.65
40	1.68	2.02	2.70	3.55
60	1.67	2.00	2.66	3.46
120	1.66	1.98	2.62	3.37
∞	1.64	1.96	2.58	3.29

$p > 0.05$	$p < 0.05$	$p < 0.01$	$p < 0.001$
Not significant (NS)	Significant	Highly significant	Very highly significant
	(fairly confident)	(very confident)	(almost certain)

APPENDIX 3B

Graph of *t* distribution

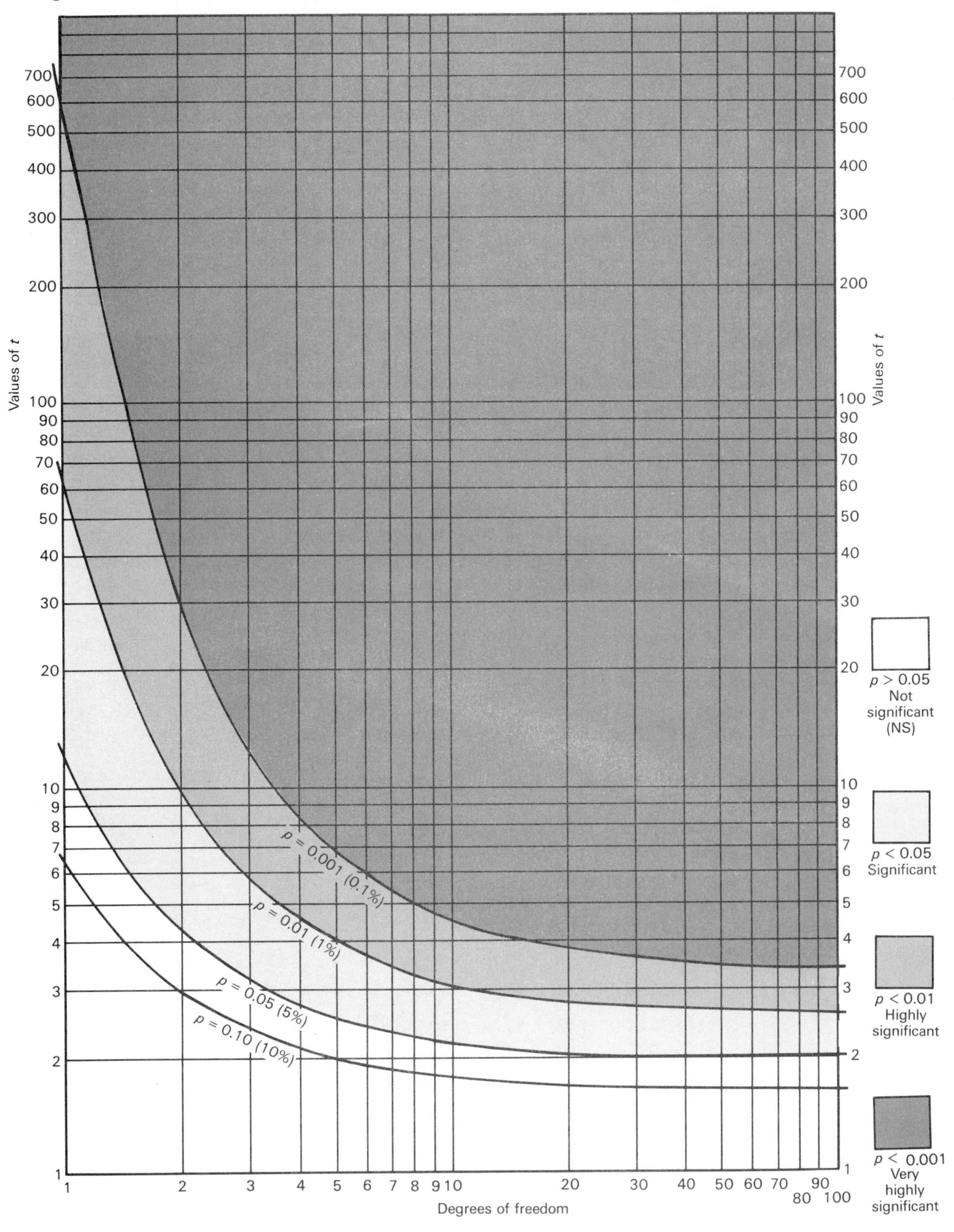

APPENDIX 4A

Table of the χ^2 distribution

Increasing value of p ← → Decreasing value of p

df	0.99	0.95	0.90	0.50	0.10	0.05	0.01	0.001	df
	p values								
1	0.000 16	0.0039	0.016	0.46	2.71	3.84	6.63	10.83	1
2	0.02	0.10	0.21	1.39	4.60	5.99	9.21	13.82	2
3	0.12	0.35	0.58	2.37	6.25	7.81	11.34	16.27	3
4	0.30	0.71	1.06	3.36	7.78	9.49	13.28	18.46	4
5	0.55	1.14	1.61	4.35	9.24	11.07	15.09	20.52	5
6	0.87	1.64	2.20	5.35	10.64	12.59	16.81	22.46	6
7	1.24	2.17	2.83	6.35	12.02	14.07	18.48	24.32	7
8	1.65	2.73	3.49	7.34	13.36	15.51	20.09	26.12	8
9	2.09	3.32	4.17	8.34	14.68	16.92	21.67	27.88	9
10	2.56	3.94	4.86	9.34	15.99	18.31	23.21	29.59	10
11	3.05	4.58	5.58	10.34	17.28	19.68	24.72	31.26	11
12	3.57	5.23	6.30	11.34	18.55	21.03	26.22	32.91	12
13	4.11	5.89	7.04	12.34	19.81	22.36	27.69	34.53	13
14	4.66	6.57	7.79	13.34	21.06	23.68	29.14	36.12	14
15	5.23	7.26	8.55	14.34	22.31	25.00	30.58	37.70	15
16	5.81	7.96	9.31	15.34	23.54	26.30	32.00	39.29	16
17	6.41	8.67	10.08	16.34	24.77	27.59	33.41	40.75	17
18	7.02	9.39	10.86	17.34	25.99	28.87	34.80	42.31	18
19	7.63	10.12	11.65	18.34	27.20	30.14	36.19	43.82	19
20	8.26	10.85	12.44	19.34	28.41	31.41	37.57	45.32	20
21	8.90	11.59	13.24	20.34	29.62	32.67	38.93	46.80	21
22	9.54	12.34	14.04	21.34	30.81	33.92	40.29	48.27	22
23	10.20	13.09	14.85	22.34	32.01	35.17	41.64	49.73	23
24	10.86	13.85	15.66	23.34	33.20	36.42	42.98	51.18	24
25	11.52	14.61	16.47	24.34	34.38	37.65	44.31	52.62	25
26	12.20	15.38	17.29	25.34	35.56	38.88	45.64	54.05	26
27	12.88	16.15	18.11	26.34	36.74	40.11	46.96	55.48	27
28	13.56	16.93	18.94	27.34	37.92	41.34	48.28	56.89	28
29	14.26	17.71	19.77	28.34	39.09	42.56	49.59	58.30	29
30	14.95	18.49	20.60	29.34	40.26	43.77	50.89	59.70	30
40	22.16	26.51	29.05	39.34	51.81	55.76	63.69	73.40	40
60	37.48	43.19	46.46	59.33	74.40	79.08	88.38	99.61	60
80	53.54	60.39	64.28	79.33	96.58	101.88	112.33	124.84	80
100	70.06	77.93	82.36	99.33	118.50	124.34	135.81	149.45	100

$p > 0.90$	$p > 0.05$	$p < 0.05$	$p < 0.01$	$p < 0.001$
Result dubious or questionable	Not significant (NS)	Significant (fairly confident)	Highly significant (very confident)	Very highly significant (almost certain)

APPENDIX 4B

Graph of the χ^2 distribution

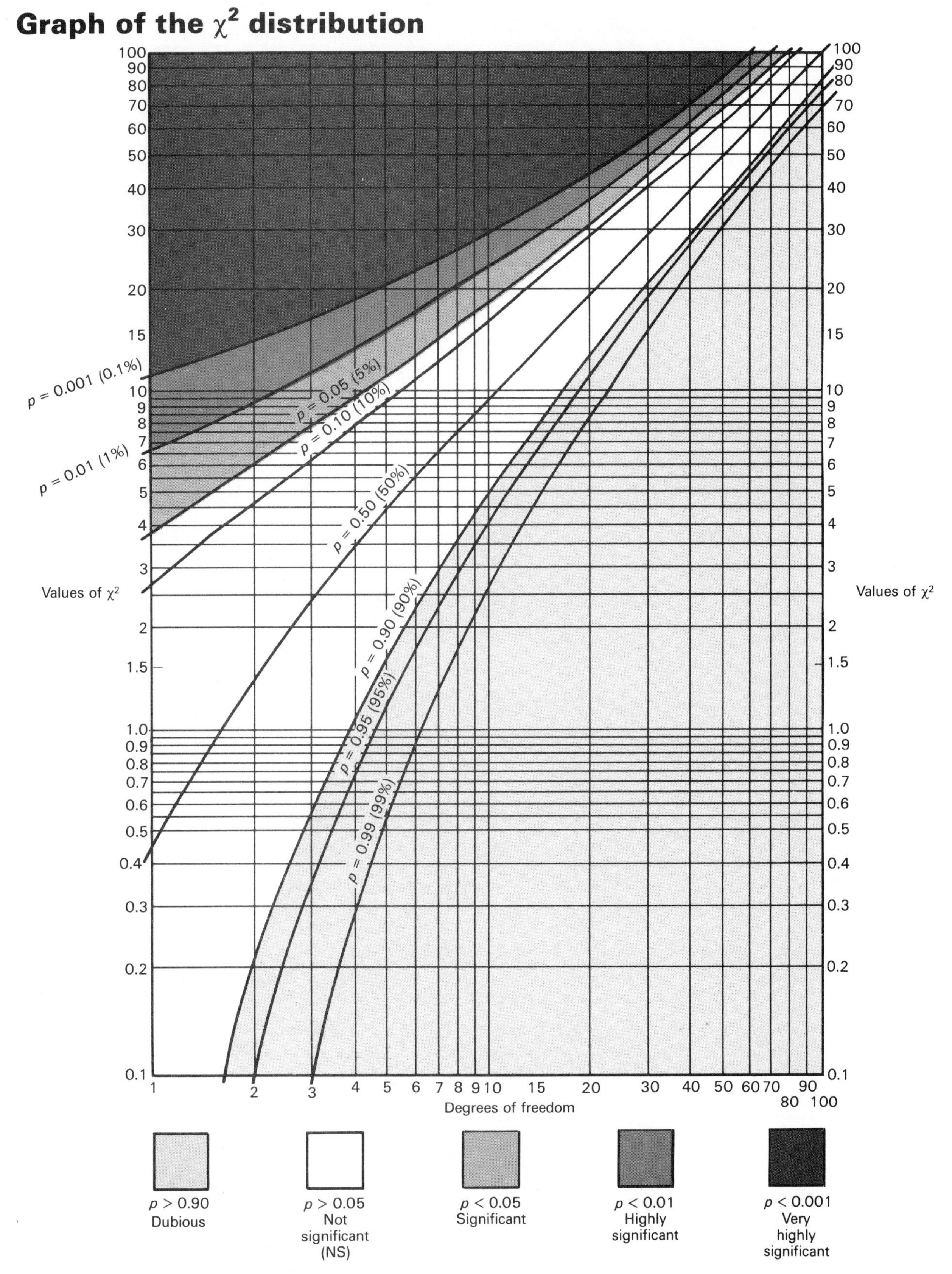

APPENDIX 5A

Table of correlation coefficient *r*

df	*p* values			
	0.1	0.05	0.01	0.001
1	0.987 69	0.996 92	0.999 877	0.999 998 8
2	0.900 00	0.950 00	0.990 000	0.999 00
3	0.8054	0.8783	0.958 73	0.991 16
4	0.7293	0.8114	0.917 20	0.974 06
5	0.6694	0.7545	0.8745	0.950 74
6	0.6215	0.7067	0.8343	0.924 93
7	0.5822	0.6664	0.7977	0.8982
8	0.5494	0.6319	0.7646	0.8721
9	0.5214	0.6021	0.7348	0.8471
10	0.4973	0.5760	0.7079	0.8233
11	0.4762	0.5529	0.6835	0.8010
12	0.4575	0.5324	0.6614	0.7800
13	0.4409	0.5139	0.6411	0.7603
14	0.4259	0.4973	0.6226	0.7420
15	0.4124	0.4821	0.6055	0.7246
16	0.4000	0.4683	0.5897	0.7084
17	0.3887	0.4555	0.5751	0.6932
18	0.3783	0.4438	0.5614	0.6787
19	0.3687	0.4329	0.5487	0.6652
20	0.3598	0.4227	0.5368	0.6524
25	0.3233	0.3809	0.4869	0.5974
30	0.2960	0.3494	0.4487	0.5541
35	0.2746	0.3246	0.4182	0.5189
40	0.2573	0.3044	0.3932	0.4896
45	0.2428	0.2875	0.3721	0.4648
50	0.2306	0.2732	0.3541	0.4433
60	0.2108	0.2500	0.3248	0.4078
70	0.1954	0.2319	0.3017	0.3799
80	0.1829	0.2172	0.2830	0.3568
90	0.1726	0.2050	0.2673	0.3375
100	0.1638	0.1946	0.2540	0.3211

$p > 0.05$	$p < 0.05$	$p < 0.01$	$p < 0.001$
No significant correlation	Significant correlation (fairly confident)	Highly significant correlation (very confident)	Very highly significant correlation (almost certain)

APPENDIX 5B

Graph of correlation coefficient *r*

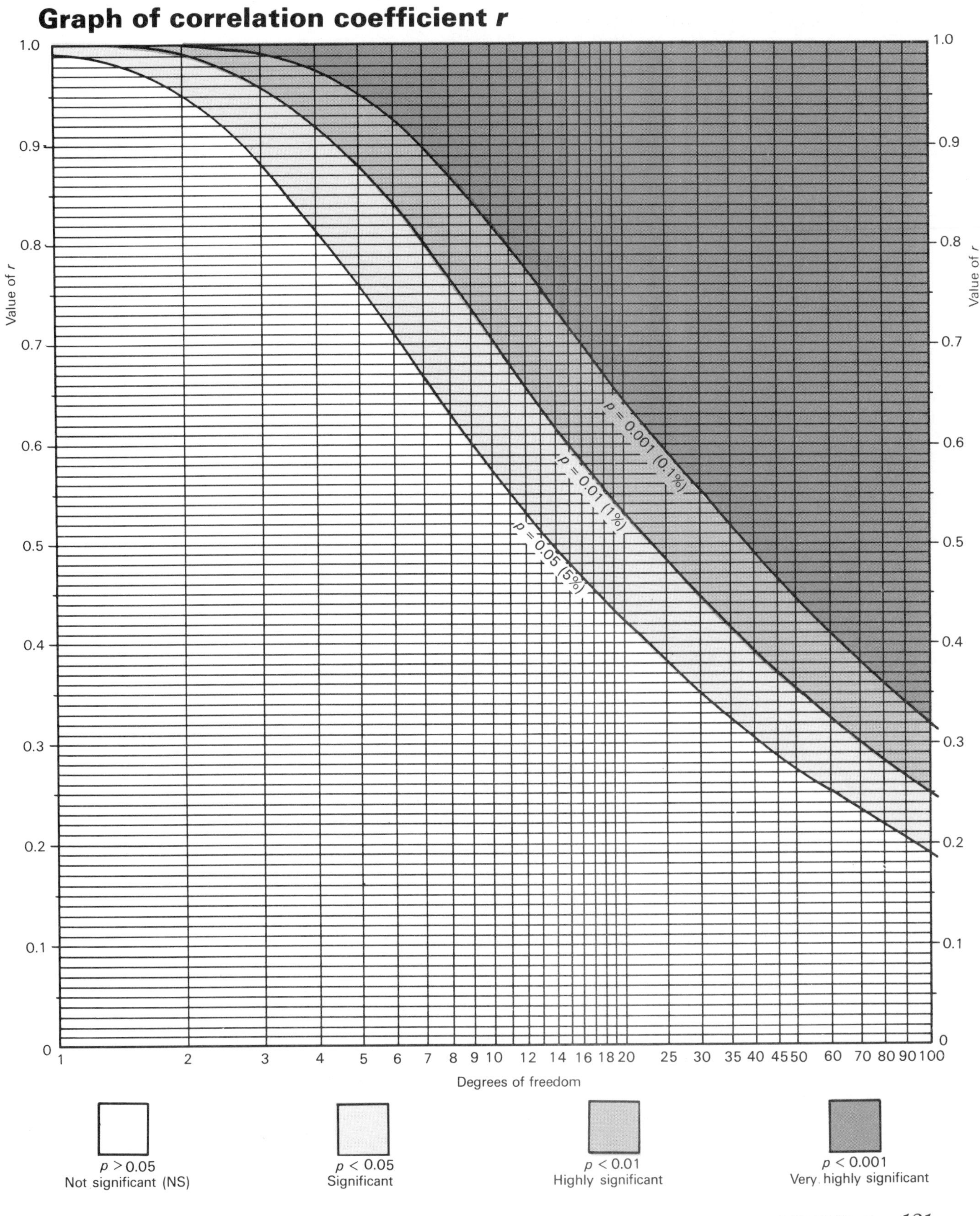

Guide

INTRODUCTION

The Guide contains answers to all the questions set in the text exercises. It also contains details of all the graphs that you have to construct plus comments.

Do not be tempted into looking at the answers before completing the questions. It is important that you understand everything that you have to do, so it is helpful and encouraging to have readily available feedback to inform you of your progress.

Work your way through a complete exercise or even a complete section before checking that you have answered correctly. If you have answered incorrectly, try to find out exactly where you went wrong and why. If necessary work backwards through the exercise; we can learn so much from our mistakes. If you still have any difficulties and don't understand the question or where you went wrong, ask for advice.

The guide is cross-referenced to the text sections and page numbers and also to the Work Sheet numbers.

1 DATA

1.2.3, p.3

1 Discrete

2 Continuous

3 Qualitative

4 Continuous

You may have had some difficulty with question 4 since the data are given in whole numbers. While some variables are typically discrete, in practice all measurements are discrete variates since there are limitations of accuracy in any measurement. The degree to which measurements approach continuity will depend on the fineness with which the units of measurement can be subdivided.

1.3, p.4

1 Distance (mm)

2 Time (min)

2 ORGANISING AND PRESENTING DATA

2.1.1, p.7

1 16–18 years

2 13–15 years

3 Daily energy and protein requirements follow each other very closely. Male and female figures match each other up to about 8 years – above this age the requirements for the male exceed that of the female. Males and females are equal in mass up to about 11 years, then the girls spurt ahead for a couple of years, then the boys pass the girls and are heavier from about 16 onwards. The male values are never lower than the female values, but the mass rises to a particular level whereas the protein and energy requirements reach a peak (for the boys 16–18 years and for the girls 13–15) then drop.

The graphs below show the relationships between the variables more clearly.

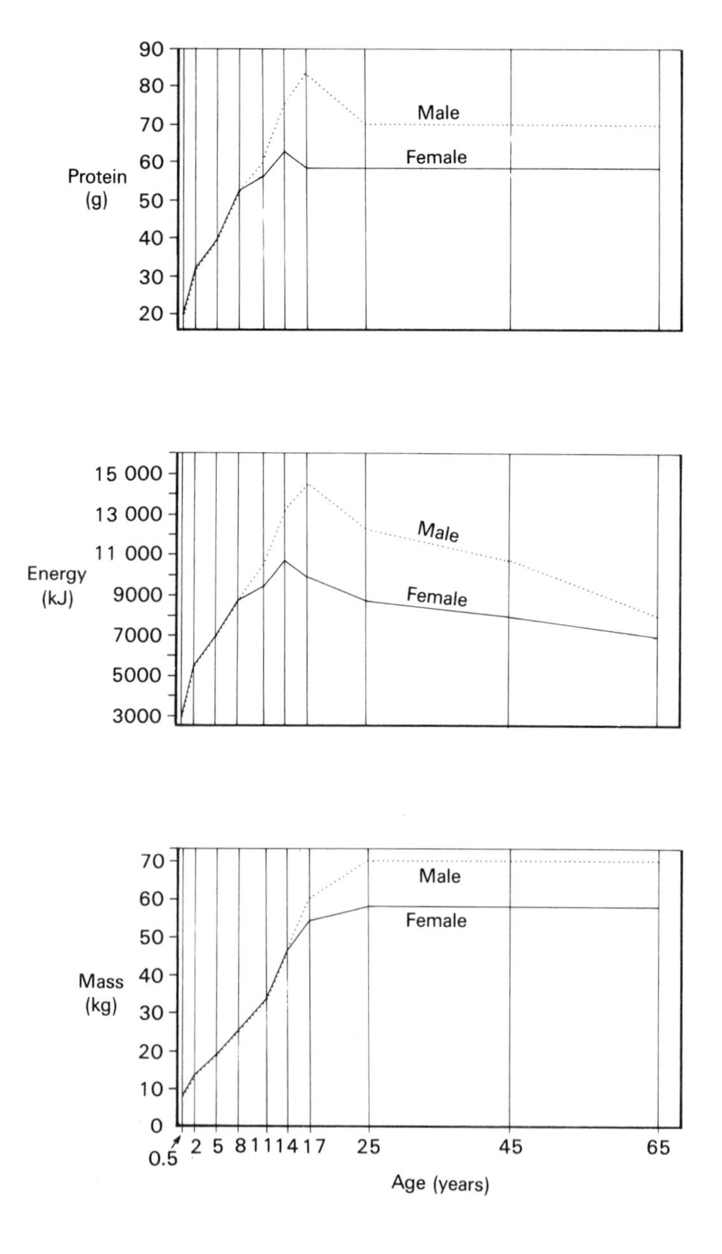

2.1.3, p.10, Work Sheet 1

1

	Tally columns						
Class	**5**	**10**	**15**	**20**	**25**	**30**	**Frequency**
4.01– 5.20	卌	IIII					9
5.21– 6.40	卌	卌	卌	卌	I		21
6.41– 7.60	卌	卌	卌	卌	卌	II	27
7.61– 8.80	卌	卌	卌				14
8.81–10.00	卌	卌	卌	II			17
10.01–11.20	卌	卌	I				11
11.21–12.40	卌	III					8
12.41–13.60	I						1
13.61–14.80	I						1
14.81–16.00	I						1

$\sum f = 110$

2.2.1, p.14, Work Sheet 2

1

0 1 2 3 4 5 6 7 8 9 10

2

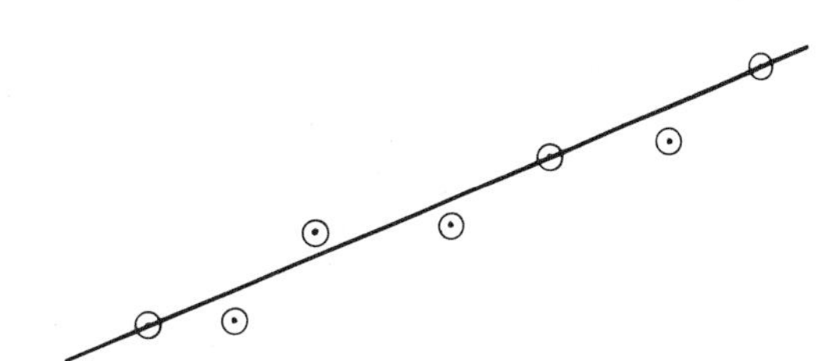

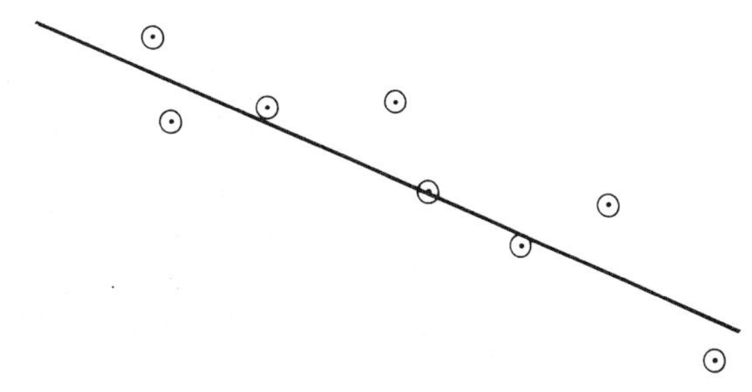

2.2.1, p.15, Work Sheet 4

1

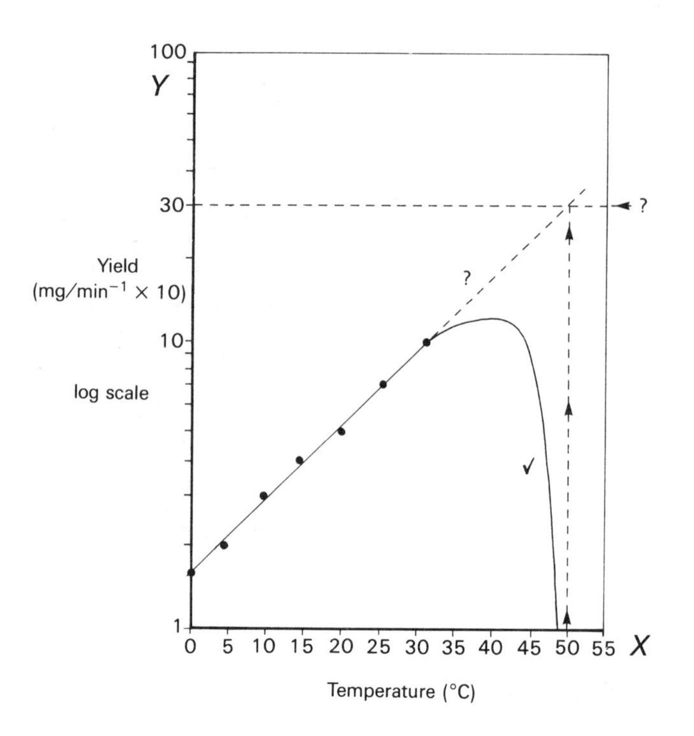

In this case you have been misled since the straight line model is only valid over a limited range of temperatures. Enzymes are denatured above 40 °C when their activity slows considerably. The yield at 50 °C would not be around 30 mg/min × 10, but around zero.

2.2.3, p.17, Work Sheet 5

1

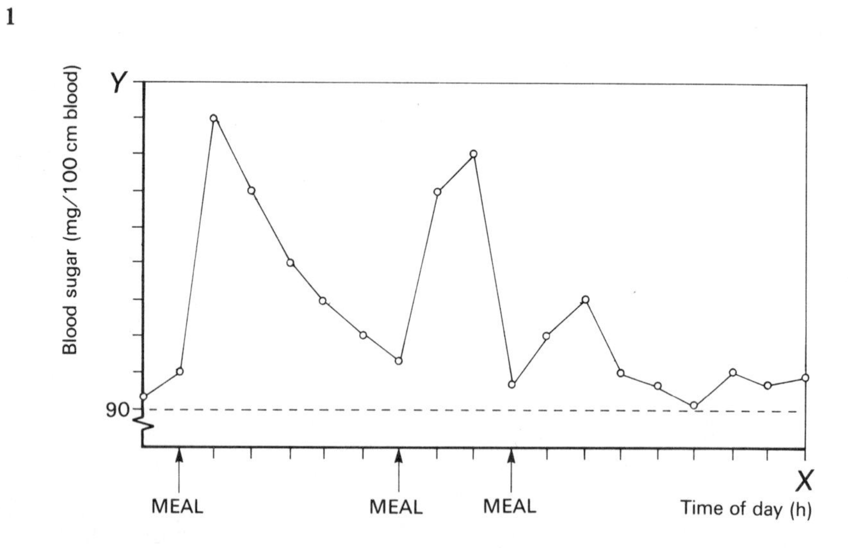

2 90 mg/100 cm^3

3 It rises due to sugar (result of carbohydrate digestion) passing into the blood from the small intestine.

4 Sugar is used up during respiration to produce energy; and stored as glycogen in the liver.

2.2.3, p.17, Work Sheet 6

1

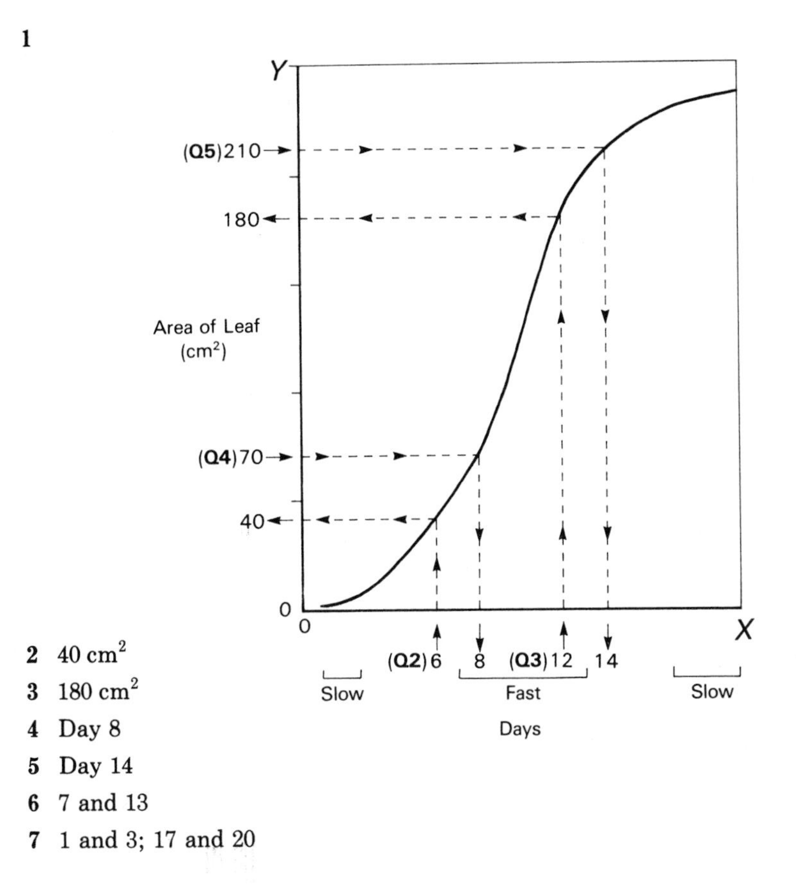

2 40 cm^2

3 180 cm^2

4 Day 8

5 Day 14

6 7 and 13

7 1 and 3; 17 and 20

2.2.3, p.18, Work Sheet 7

1

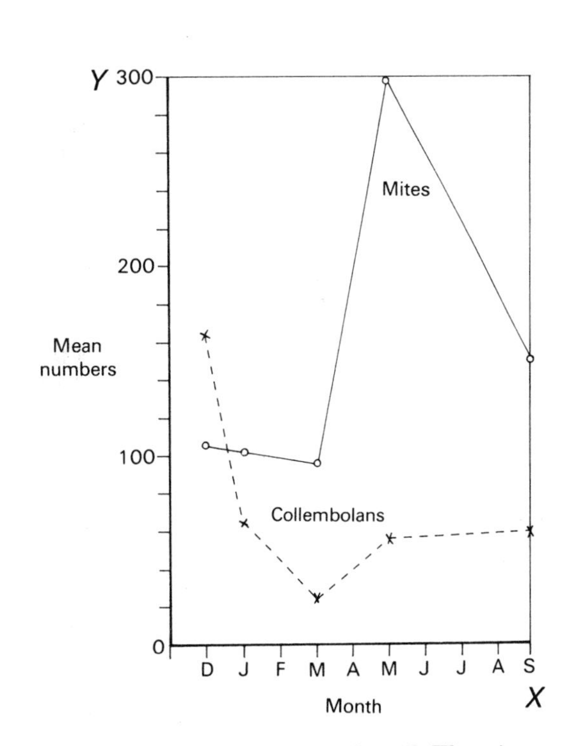

2 Both the mites and the collembolans live in the soil. The mites are predatory, living on other animals, such as the collembolans. Note that the collembolan numbers drop from January to March, then the mite numbers increase rapidly from March to June.

2.2.3, p.18, Work Sheet 8

1

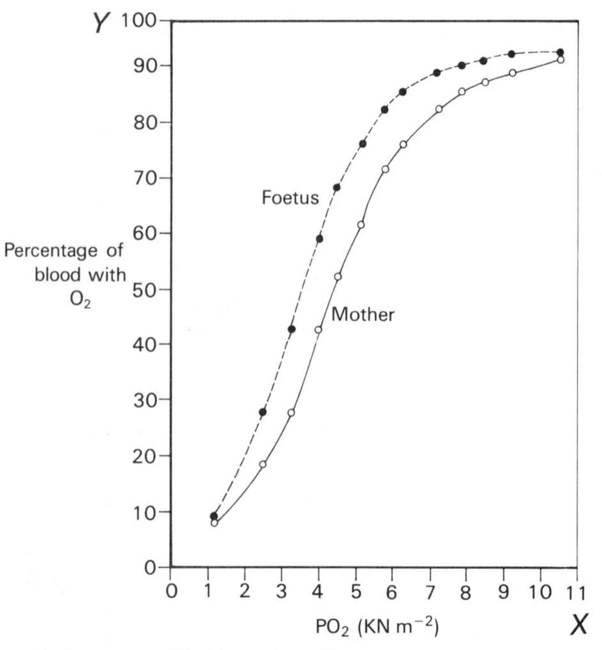

2 The two graphs have roughly the same shape.

3 The graph for the foetus lies to the left of that for the mother, indicating that the foetal blood has a greater affinity for oxygen. Thus oxygen will pass across the placenta into the foetal blood.

2.2.3, p.18, Work Sheet 9

1

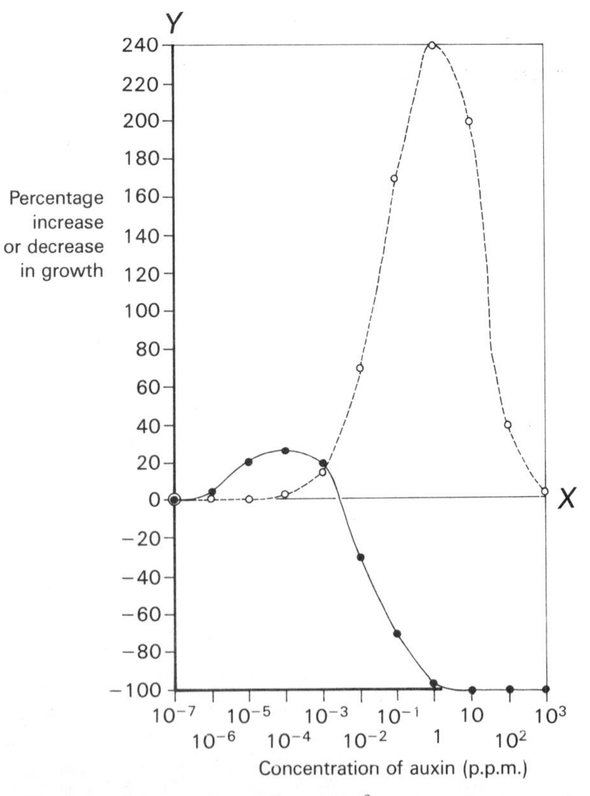

2 *Shoot:* At concentrations greater than 10^{-3} p.p.m. the growth of the shoot increases rapidly up to a concentration of about 1 p.p.m., above which growth is inhibited.

Root: At concentrations from 10^{-6} to 10^{-4} p.p.m. growth is stimulated, but above 10^{-4} p.p.m. growth is inhibited.

2.2.3, p.18, Work Sheet 10

1

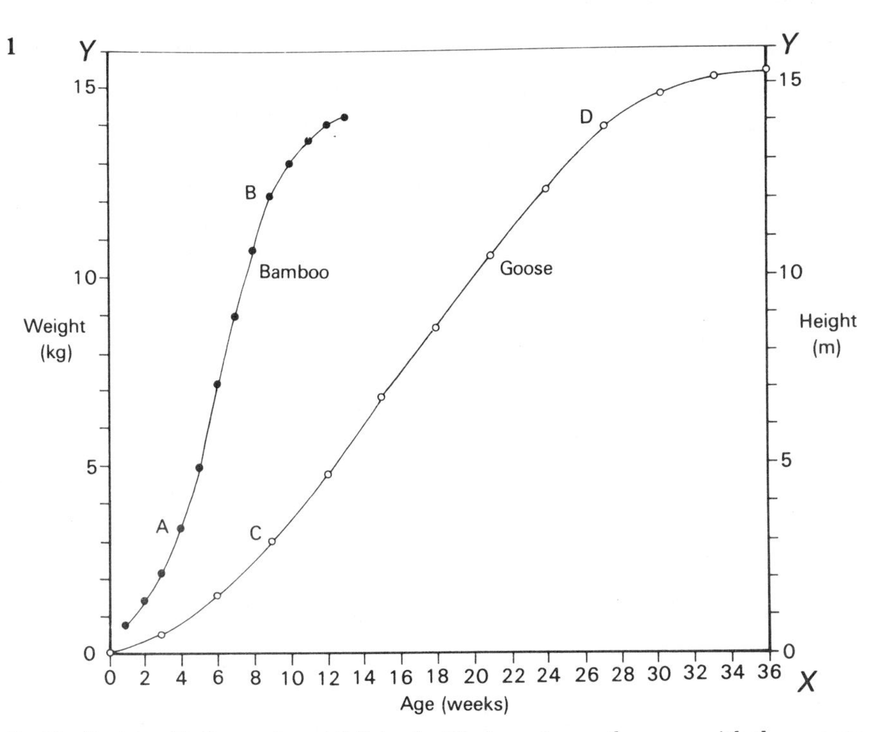

2 *Similarities:* Both graphs exhibit typical S-shaped growth curves with three stages of growth:

(i) Slow start – from the origin to A and C on the graphs.

(ii) Fast growth – A to B, and C to D on the graphs. Note both graph lines are straight during this stage; the relationship with time is proportional and growth is therefore constant.

(iii) Slowing down – from B and D on the graphs onwards.

Differences: Although the two *Y* scales are different, it would appear that the bamboo grows more rapidly than the goose, the bamboo reaching its full size after only 14 weeks, whereas the goose takes 34 weeks.

2.2.3, p.19, Work Sheet 11

1

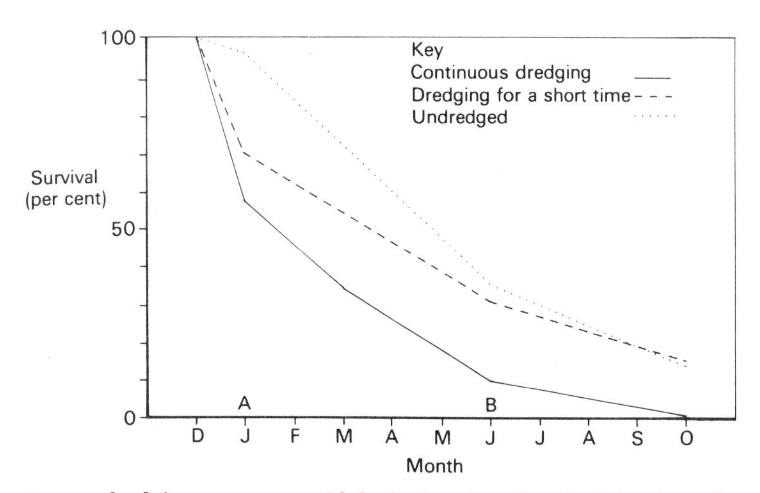

2 Continuous dredging causes a rapid depletion of stocks; dredging for a short time reduces the stocks less rapidly than continuous dredging but more rapidly than no dredging. Note that between January and June (A and B) the decrease is approximately linear in all three cases. Also note that by September or October there is no difference in the percentage survival between dredging for a short time and no dredging.

3 Dredging should be carried out only for a short time, to allow the cockles to recover by reproduction.

2.2.3, p.19, Work Sheet 12

1

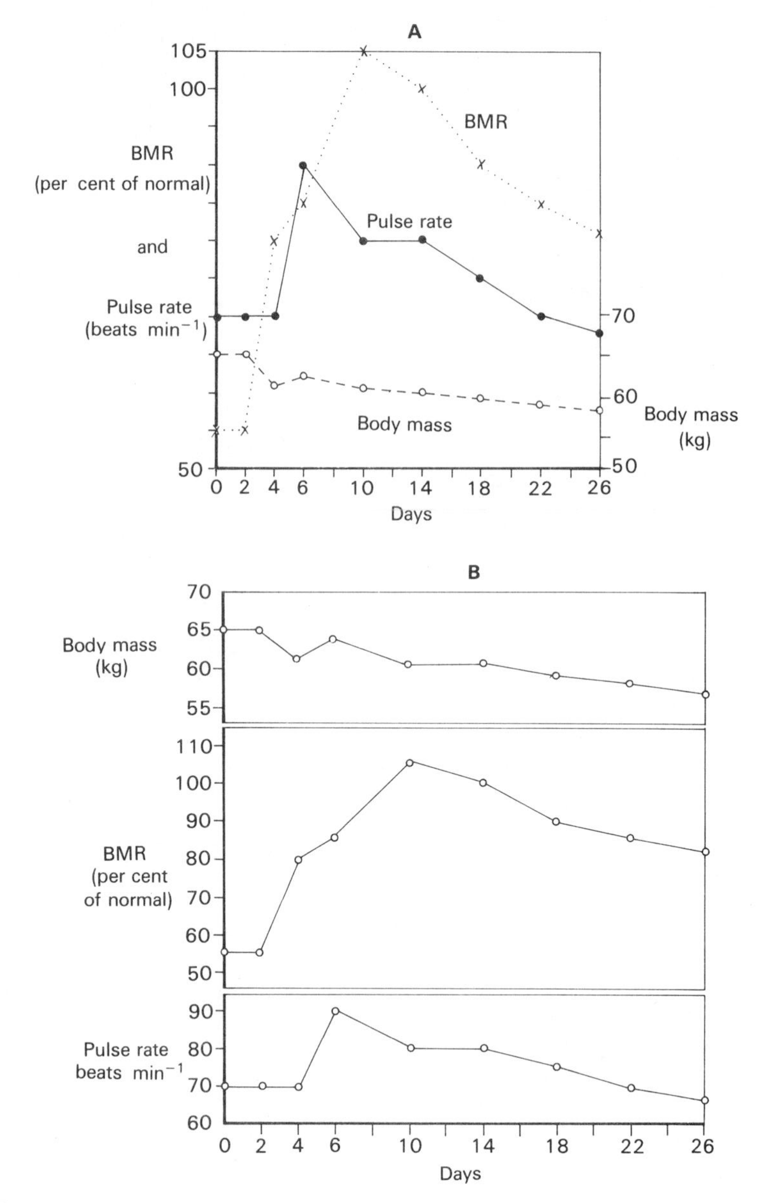

2 The placing of the graphs one above the other makes for easier comparison; the composite graph can be confusing, particularly when the graph lines cross.

2.2.3, p.22, Work Sheet 13

1 5,100,0;
5,300,75;
5,800,110;
5,1300,110;
12,100,0;
12,300,75;
12,800,110;
12,1300,110.

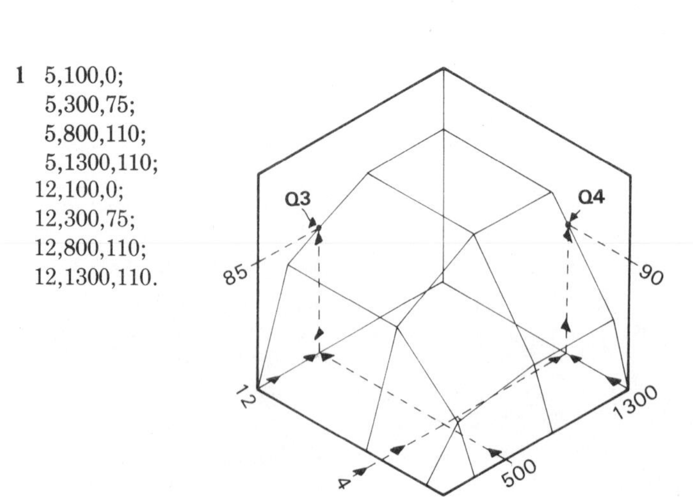

2 As the light intensity increases up to 5 units the rate of photosynthesis increases, but there is no further increase above 5 units.
As the CO_2 concentration increases up to 800 p.p.m. the rate of photosynthesis increases, but beyond this level of CO_2 there is no increase in the photosynthetic rate.

3 85 $mm^3\,cm^{-2}\,h^{-1}$

4 90 $mm^3\,cm^{-2}\,h^{-1}$

2.2.4, p.24

1 Two cycles

2 Four cycles

3 Log two cycle × log four cycle

2.2.4, p.25, Work Sheet 14

1

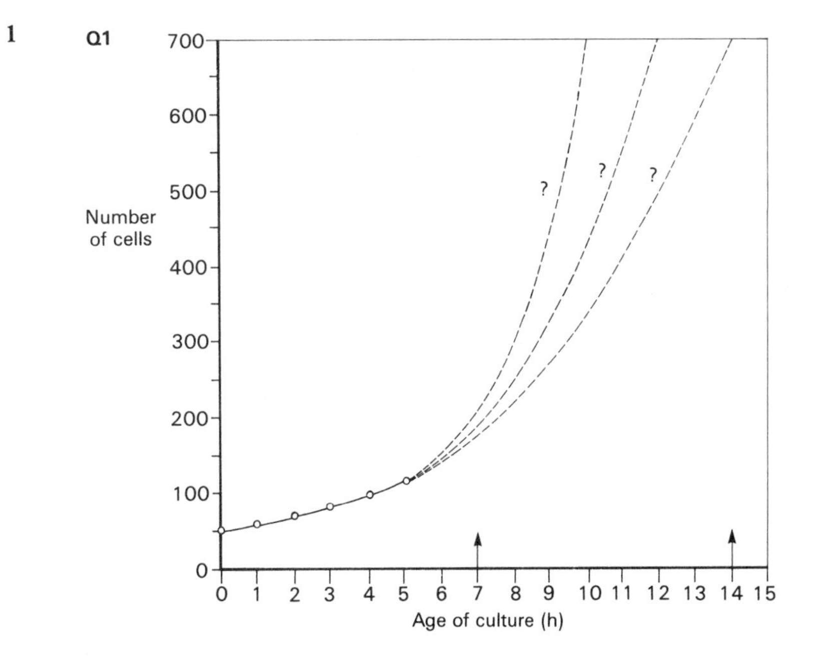

2 7 h : 180–220
14 h : 700 or greater
As you can see, extrapolation is very difficult under these circumstances. This means that estimates beyond the plotted points are virtually impossible and certainly unreliable.

2.2.4, p.25, Work Sheet 15

3

Number of cells	Log number of cells
50	1.70
60	1.78
72	1.86
86.4	1.94
103.6	2.00
124.4	2.10

4

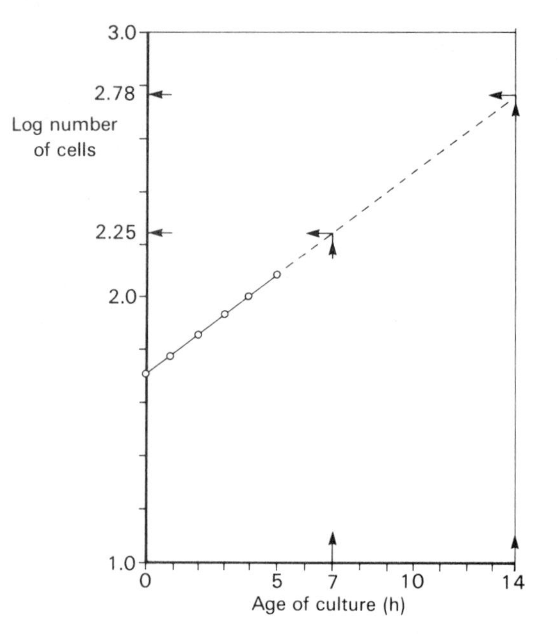

5 A straight line

6 7 h : log number of cells is 2.25, so the actual number of cells is $10^{2.25}$ = 178.
14 h : log number of cells is 2.76, so the actual number of cells is $10^{2.76}$ = 575.

7

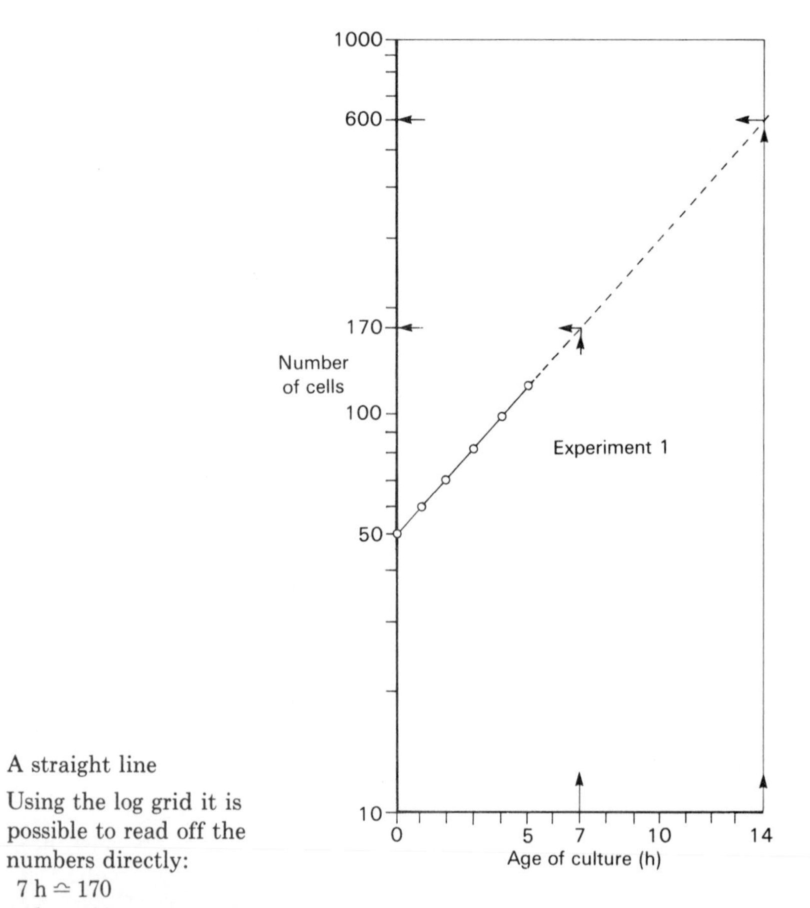

8 A straight line

9 Using the log grid it is possible to read off the numbers directly:
7 h ≏ 170
14 h ≏ 600

10 Using log grids can save a lot of time and effort. You can plot quickly and accurately without having to determine the logs of the values. When extrapolating, the population could be estimated after many more hours since the log scale compressed the larger values. A similar problem of dealing with very small values is also overcome by the log scale 'expanding' them. Extrapolation was much more straightforward since the line was straight.

2.2.4, p.25, Work Sheet 16

1

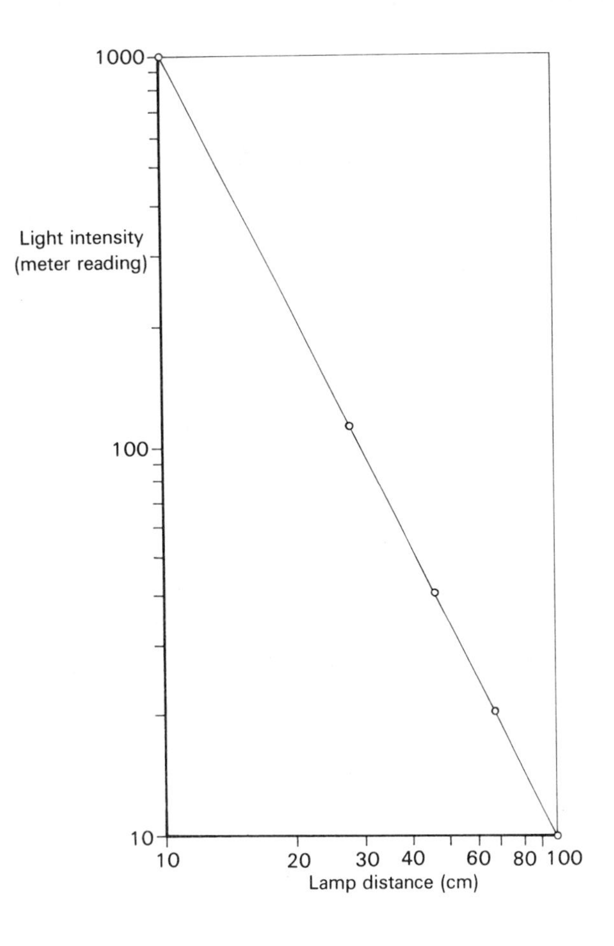

2.2.5, p.27, Work Sheet 17

1

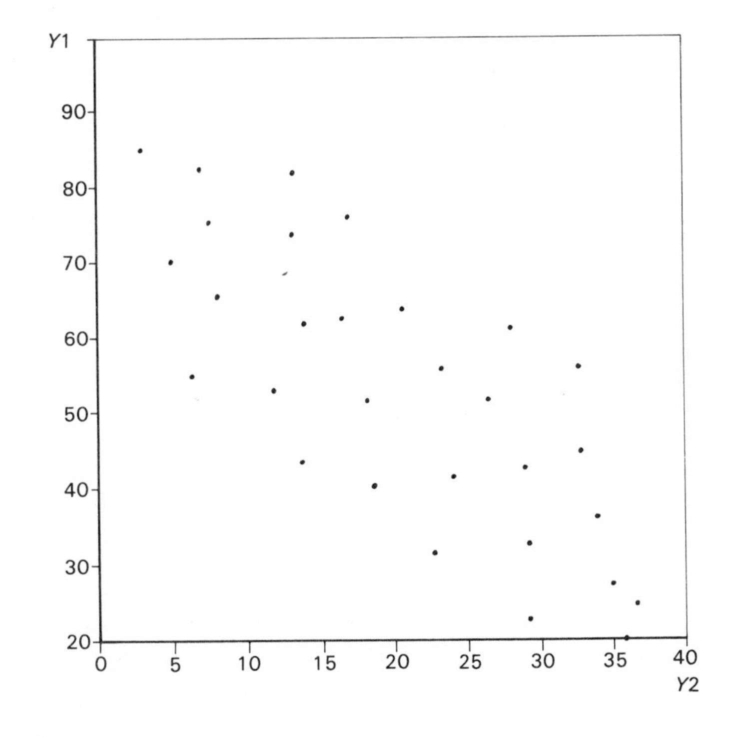

2.2.5, p.28, Work Sheet 18

1

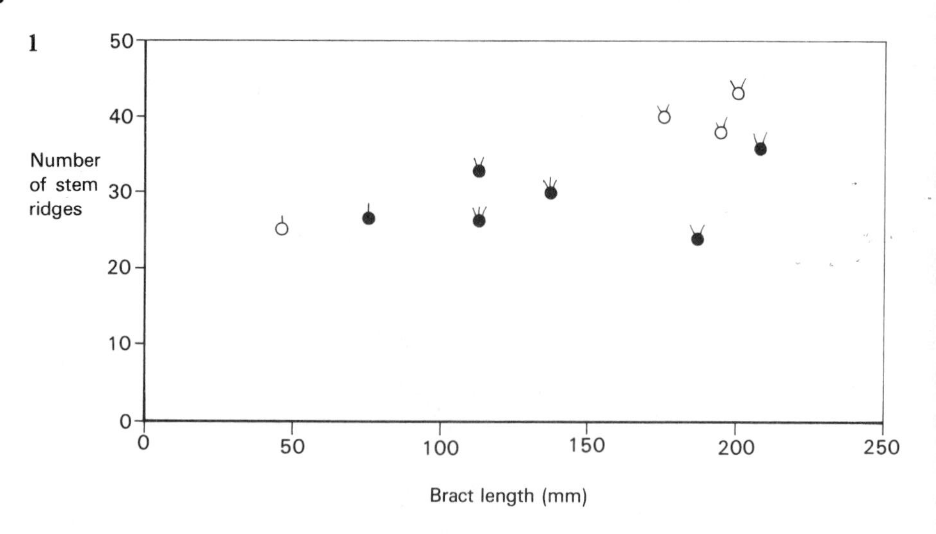

2 Yes; instead of two distinct populations, as seen in the original graph, there do not appear to be any distinct groups.

2.2.5, p.29, Work Sheet 19

1 2

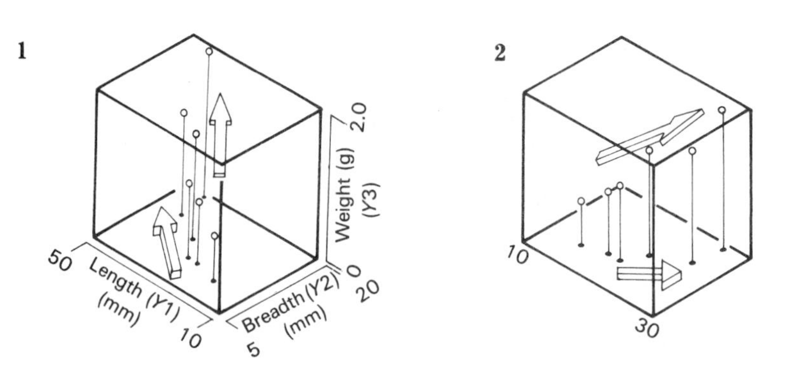

The arrows indicate correlation, i.e. it exists in two directions.

2.2.6, p.30, Work Sheet 20

1 and 2

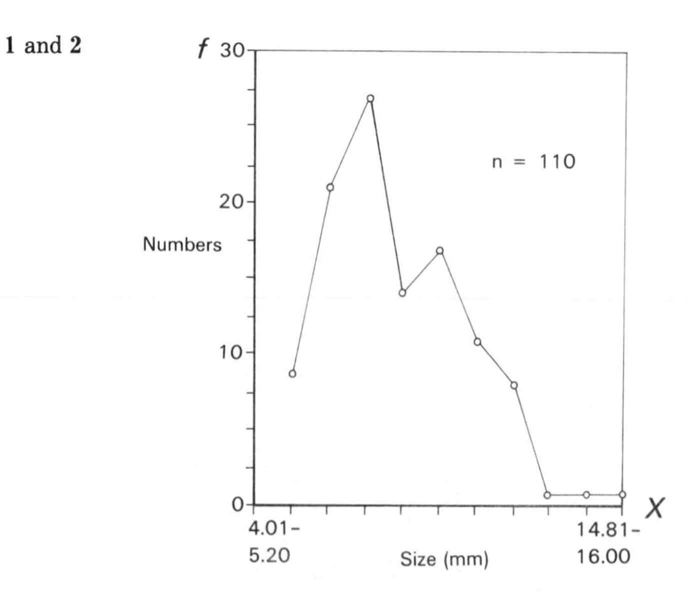

2.2.6, p.31, Work Sheet 21

1

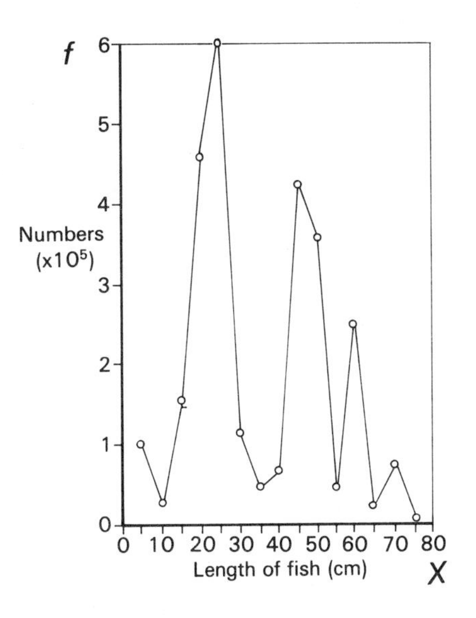

2 The four peaks stand out clearly, and probably represent four age groups, i.e. first, second, third and fourth years of life. As the fish get bigger (older) there are fewer and fewer of them. This is probably due to the heavy fishing after the war.

2.2.6, p.31, Work Sheet 22

1

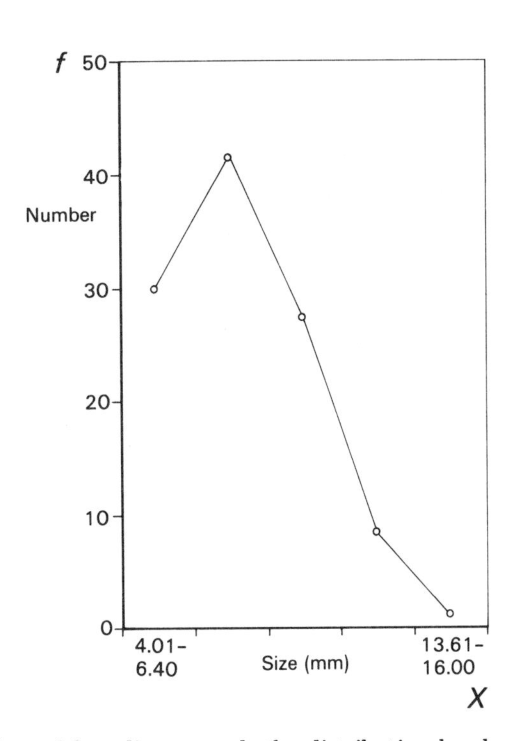

2 The second 'hump' has disappeared; the distribution has been smoothed out. The smaller number of classes can obscure important aspects of the distribution.

2.2.6, p.32, Work Sheet 23

1

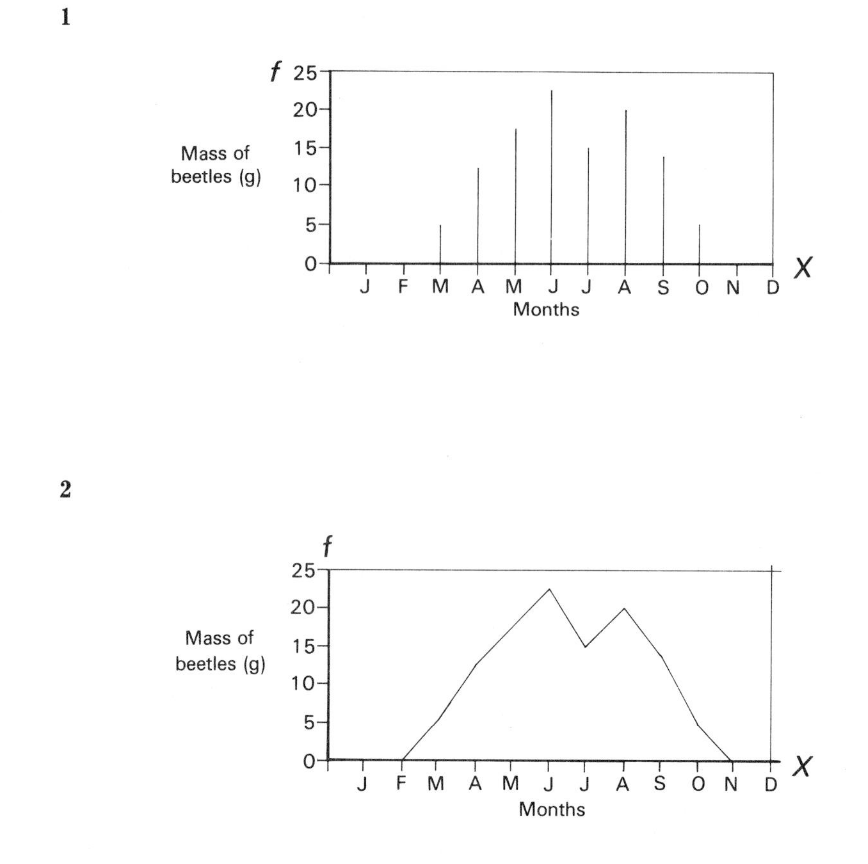

2.2.6, p.32, Work Sheet 24

1

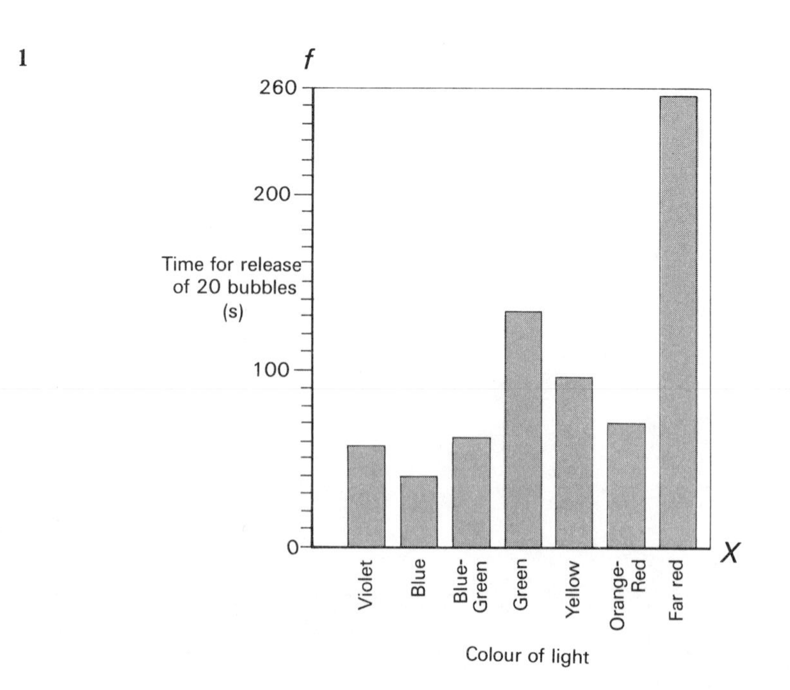

2.2.6, p.32, Work Sheet 25

1

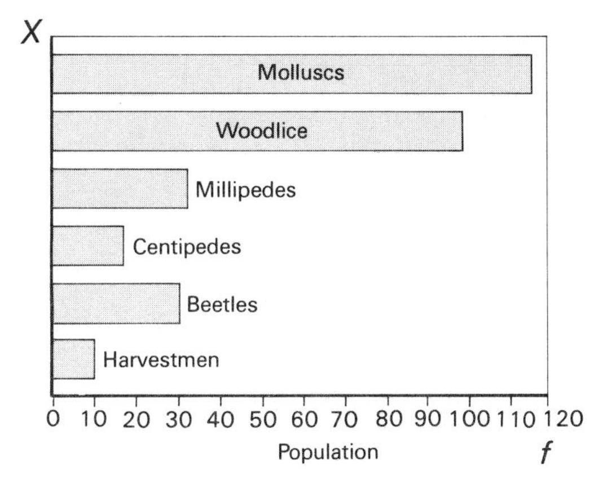

2.2.6, p.32, Work Sheet 26

1

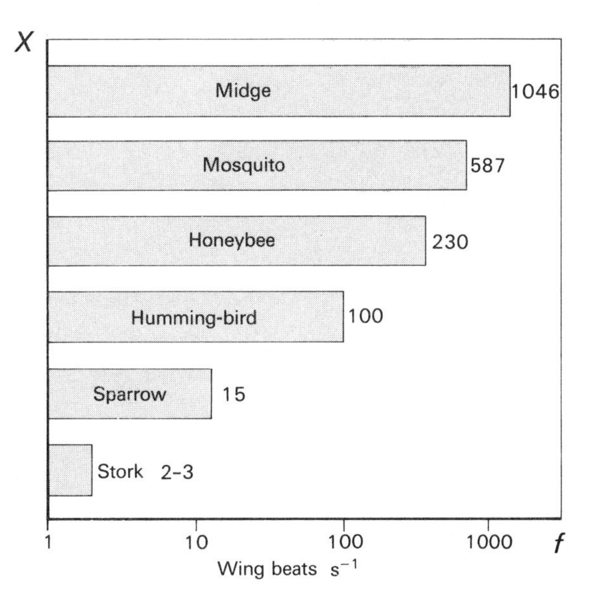

2.2.6, p.33, Work Sheet 27

1

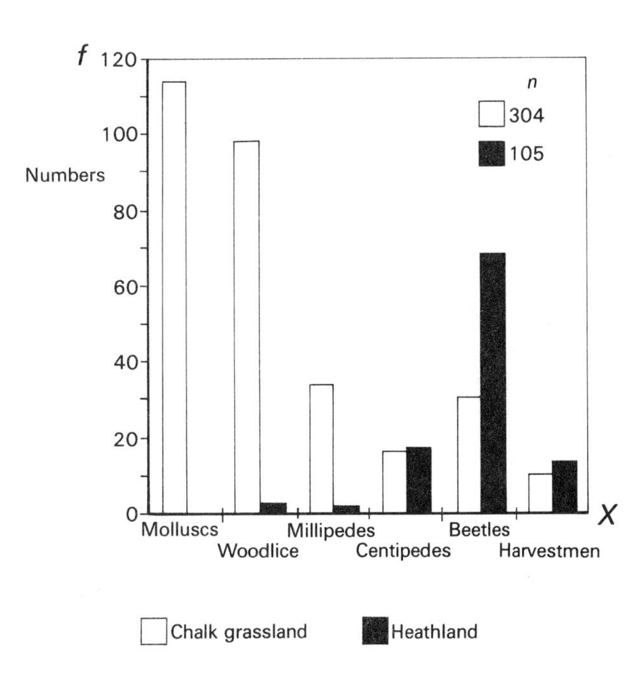

2.2.6, p.33, Work Sheet 28

1

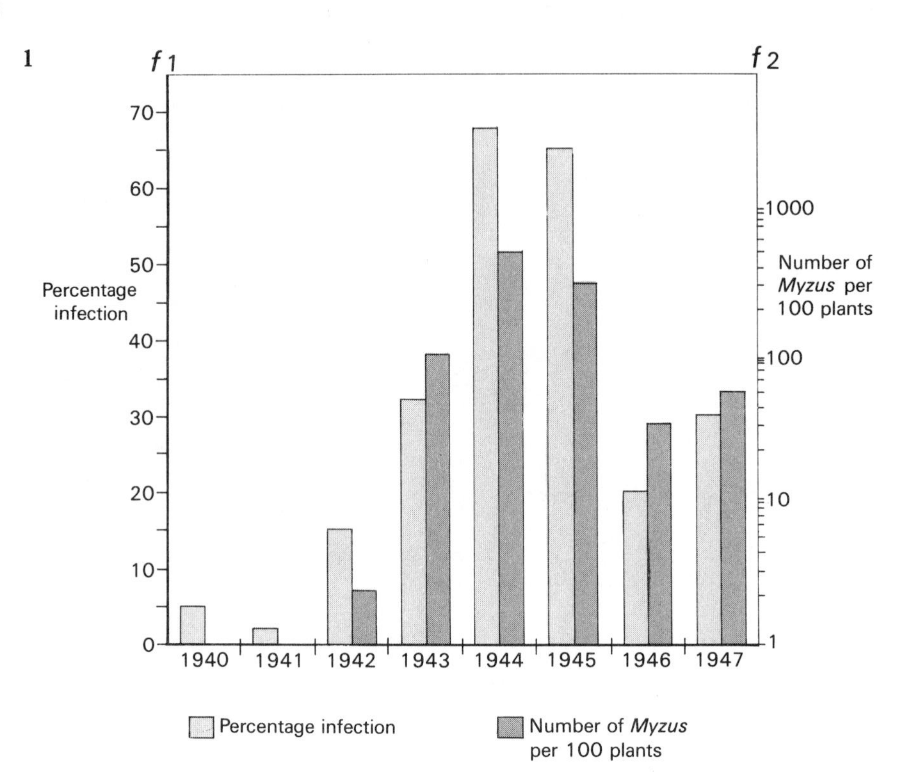

2 Yes: the higher the percentage infection the higher the numbers of *Myzus* per 100 plants.

2.2.6, p.33, Work Sheet 29

1

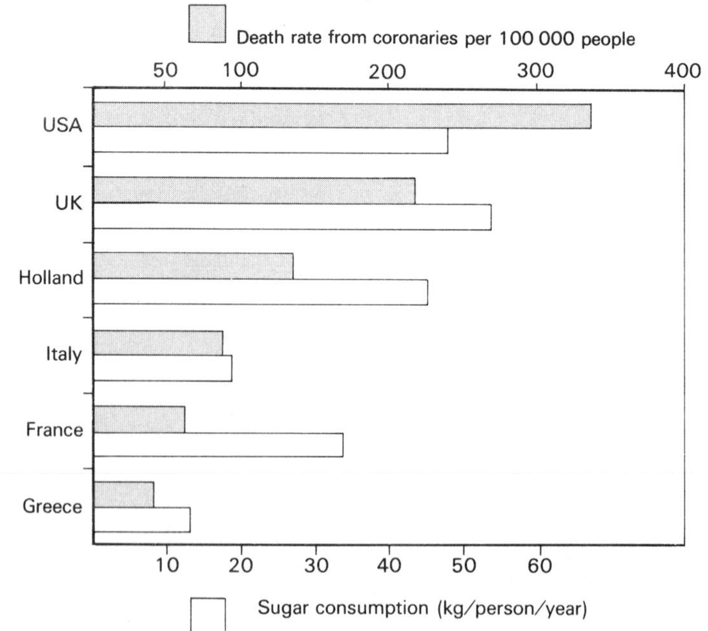

2 Yes: the higher the sugar consumption the higher the death rate from coronaries.

2.2.6, p.34, Work Sheet 30

1

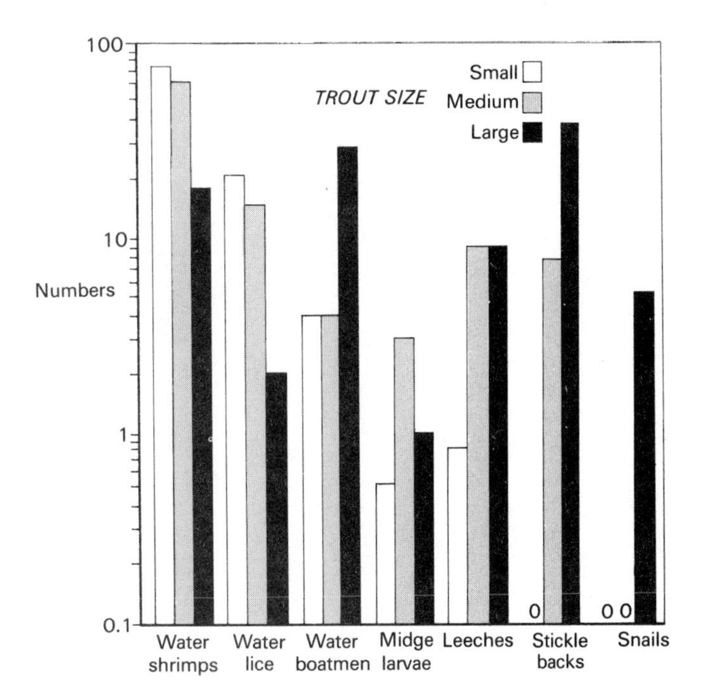

2.2.6, p.37, Work Sheet 31

1

Class	Tally columns						*f*
	5	10	15	20	25	30	
31– 40	卌	|					6
41– 50	卌	|||					8
51– 60	卌	卌	|				11
61– 70	卌	||||					9
71– 80	卌	卌	卌	卌	卌		24
81– 0	卌	卌	卌				20
1–100	卌	卌	卌	卌	卌	卌	30
101–110	卌	卌	卌	|			16
111–120	|						1
						Total	125

2

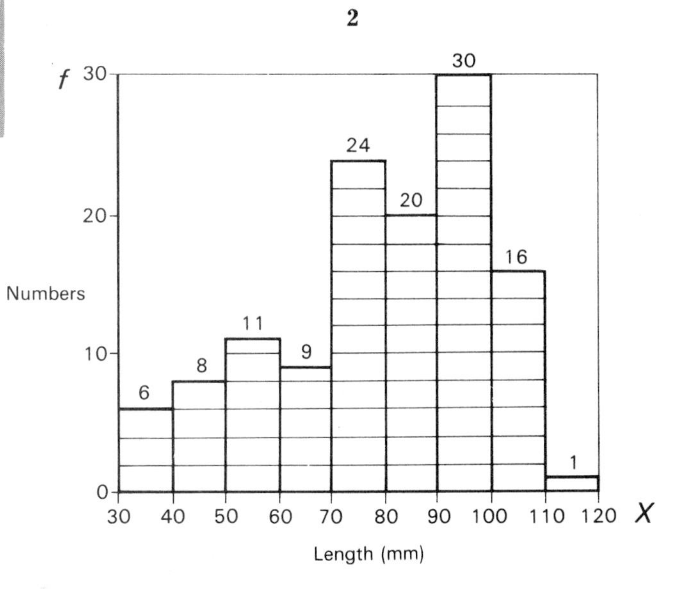

2.2.6, p.39, Work Sheet 32

1

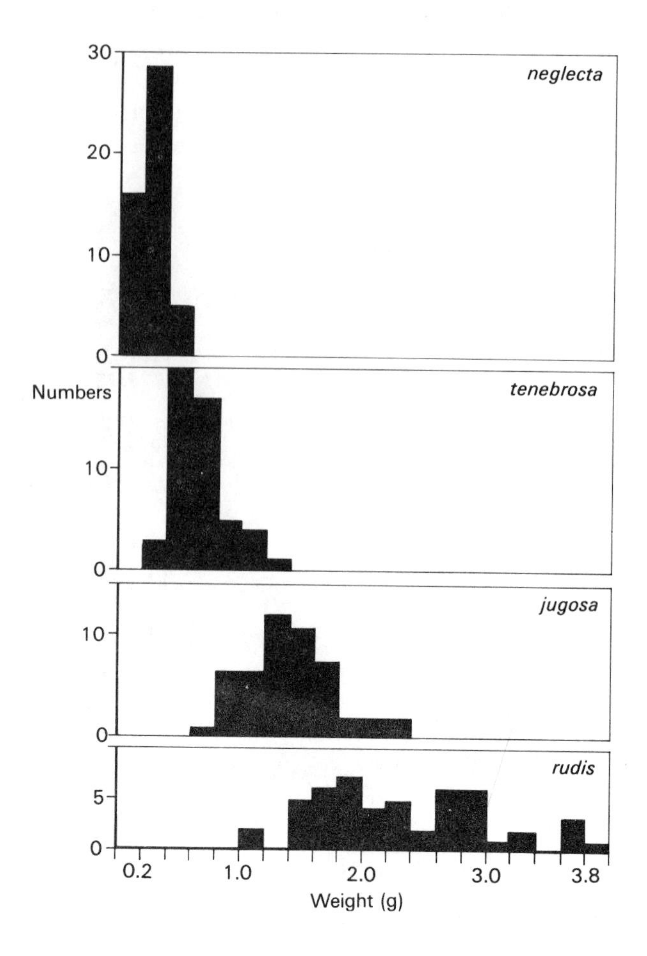

2.2.6, p.40, Work Sheet 33

1

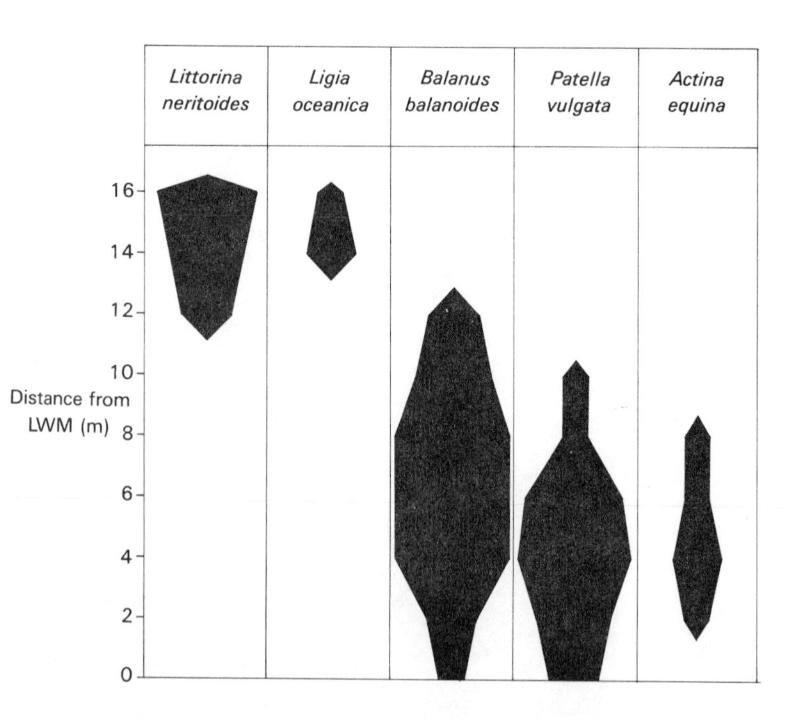

2 Yes. Although there is considerable overlap between the different species they do occupy specific zones on the shore.

2.2.6, p.41, Work Sheet 34

1

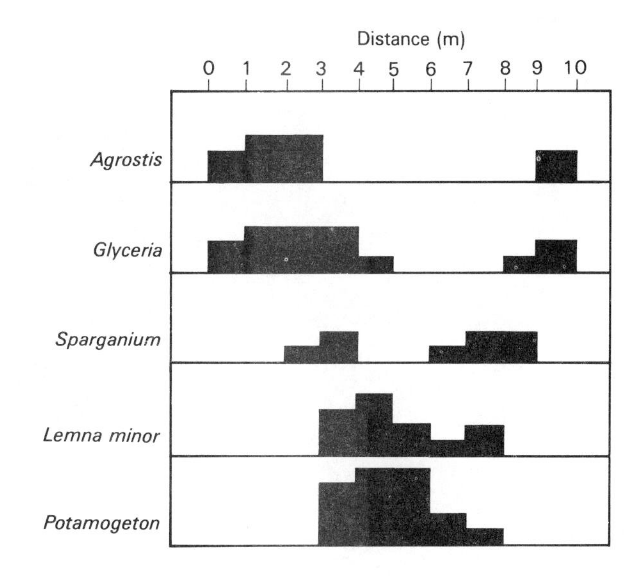

2 Yes.

2.2.7, p.43, Work Sheet 35

1

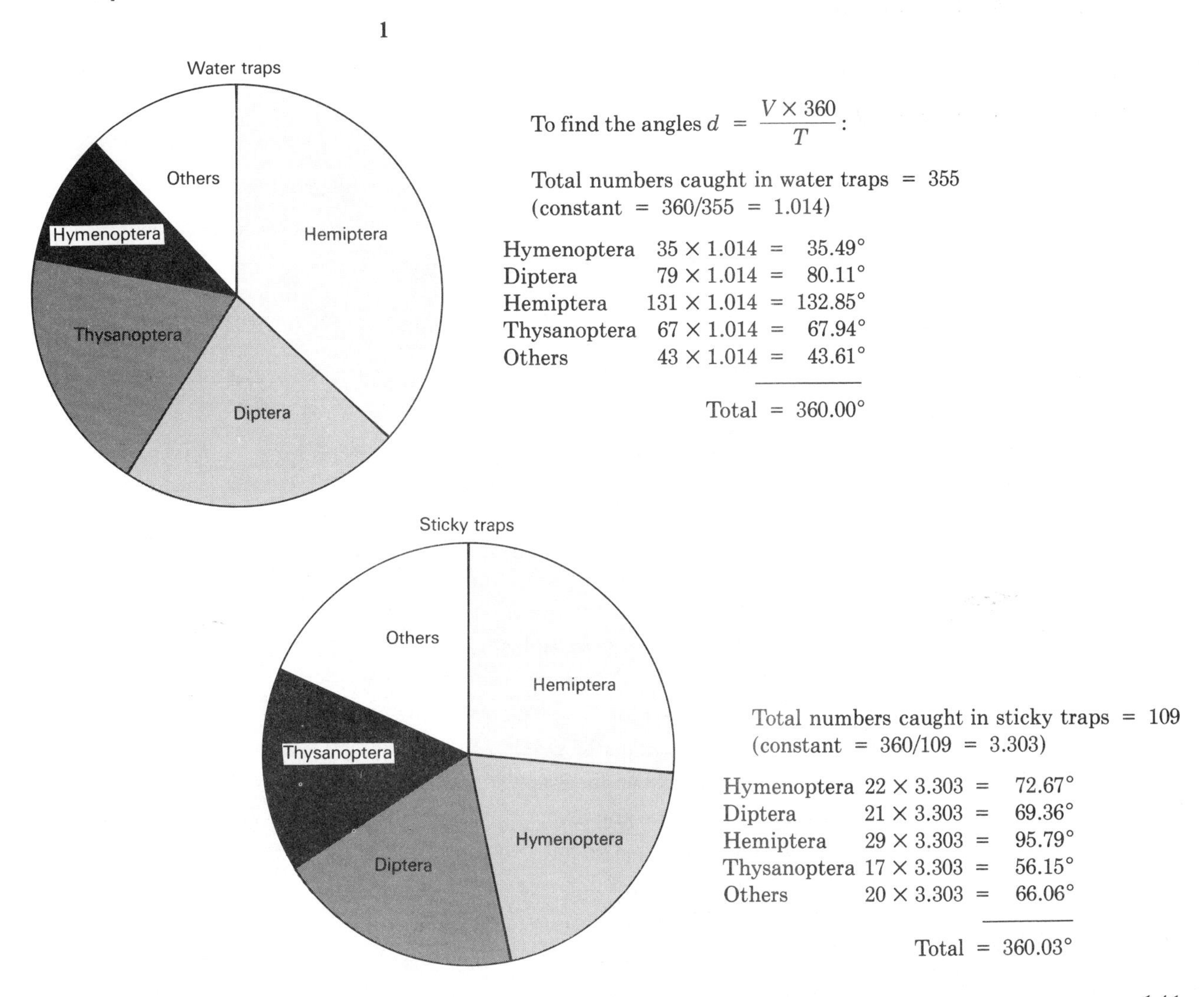

To find the angles $d = \dfrac{V \times 360}{T}$:

Total numbers caught in water traps = 355
(constant = 360/355 = 1.014)

Hymenoptera	35 × 1.014	=	35.49°
Diptera	79 × 1.014	=	80.11°
Hemiptera	131 × 1.014	=	132.85°
Thysanoptera	67 × 1.014	=	67.94°
Others	43 × 1.014	=	43.61°
	Total	=	360.00°

Total numbers caught in sticky traps = 109
(constant = 360/109 = 3.303)

Hymenoptera	22 × 3.303	=	72.67°
Diptera	21 × 3.303	=	69.36°
Hemiptera	29 × 3.303	=	95.79°
Thysanoptera	17 × 3.303	=	56.15°
Others	20 × 3.303	=	66.06°
	Total	=	360.03°

2.2.7, p.47, Work Sheet 36

1

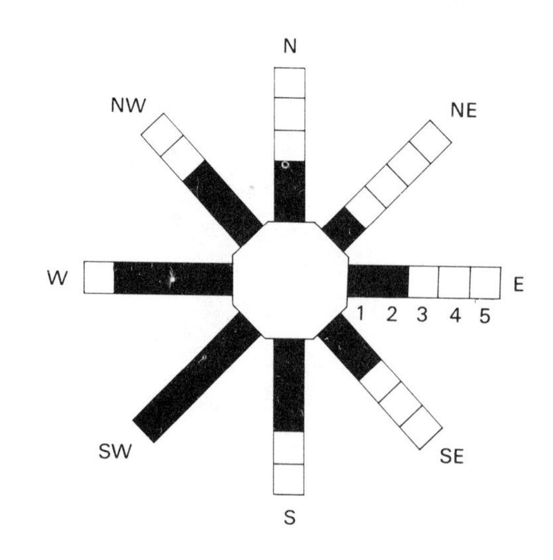

2 South-west

3 Sunlight; moisture

3 ANALYSING DATA

3.1.1, p.50, Work Sheet 37

1

		Frequencies of the genotypes					
p	q	DD	Percent	Dd	Percent	dd	Percent
0.0	1.0	0.00	0	0.00	0	1.0	100
0.1	0.9	0.01	1	0.18	18	0.81	81
0.2	0.8	0.04	4	0.32	32	0.64	64
0.3	0.7	0.09	9	0.42	42	0.49	49
0.4	0.6	0.16	16	0.48	48	0.36	36
0.5	0.5	0.25	25	0.50	50	0.25	25
0.6	0.4	0.36	36	0.48	48	0.16	16
0.7	0.3	0.49	49	0.42	42	0.09	9
0.8	0.2	0.64	64	0.32	32	0.04	4
0.9	0.1	0.81	81	0.18	18	0.01	1
1.0	0.0	1.00	100	0.00	0	0.00	0

2 The X axis.

3 See the X axis on the graph opposite.

4 See the Y axis on the graph opposite.

5 See the line DD on the graph.

6 See the line Dd on the graph.

7 See the line dd on the graph.

8 60 per cent (see graph).

9 42.5 per cent (see graph).

10 $p = 0.5$ (see graph).

11 $q = 0.67$ (see graph).

12 $p = 0.5$ (see graph).

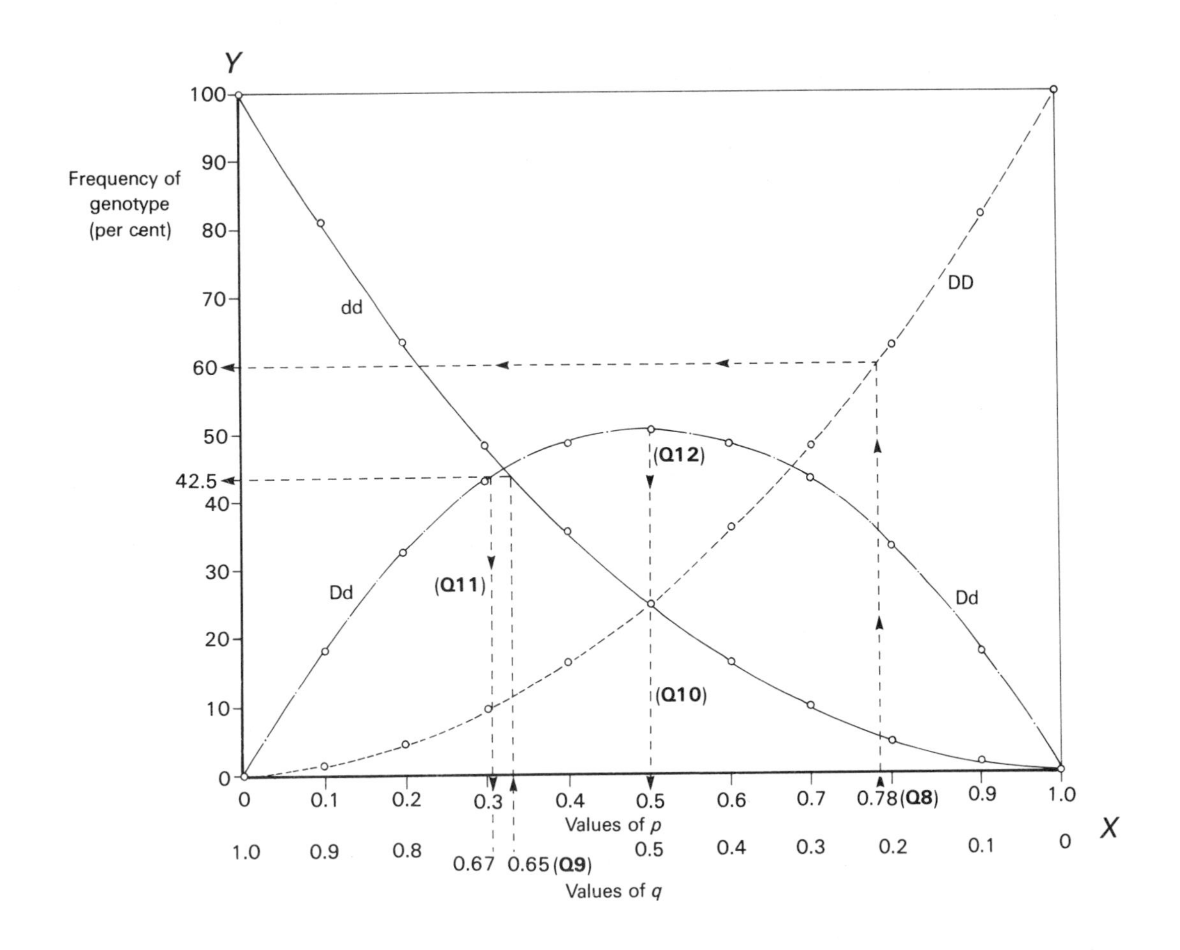

3.1.1, p.51, Work Sheet 38

1

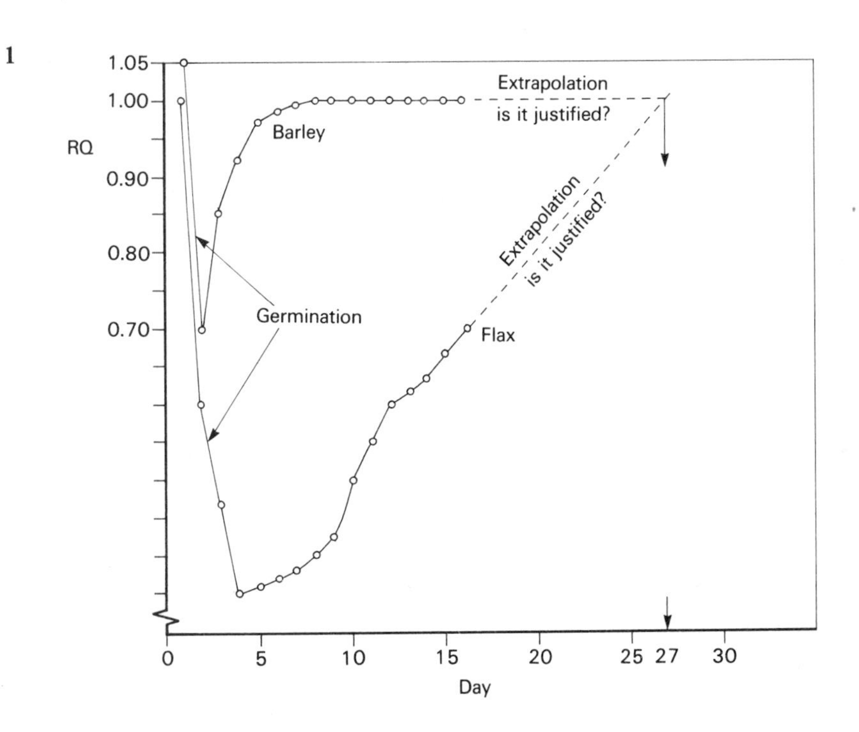

2 See graph.

3 Yes; day 27 (see graph).

3.1.1, p.51, Work Sheet 39

1

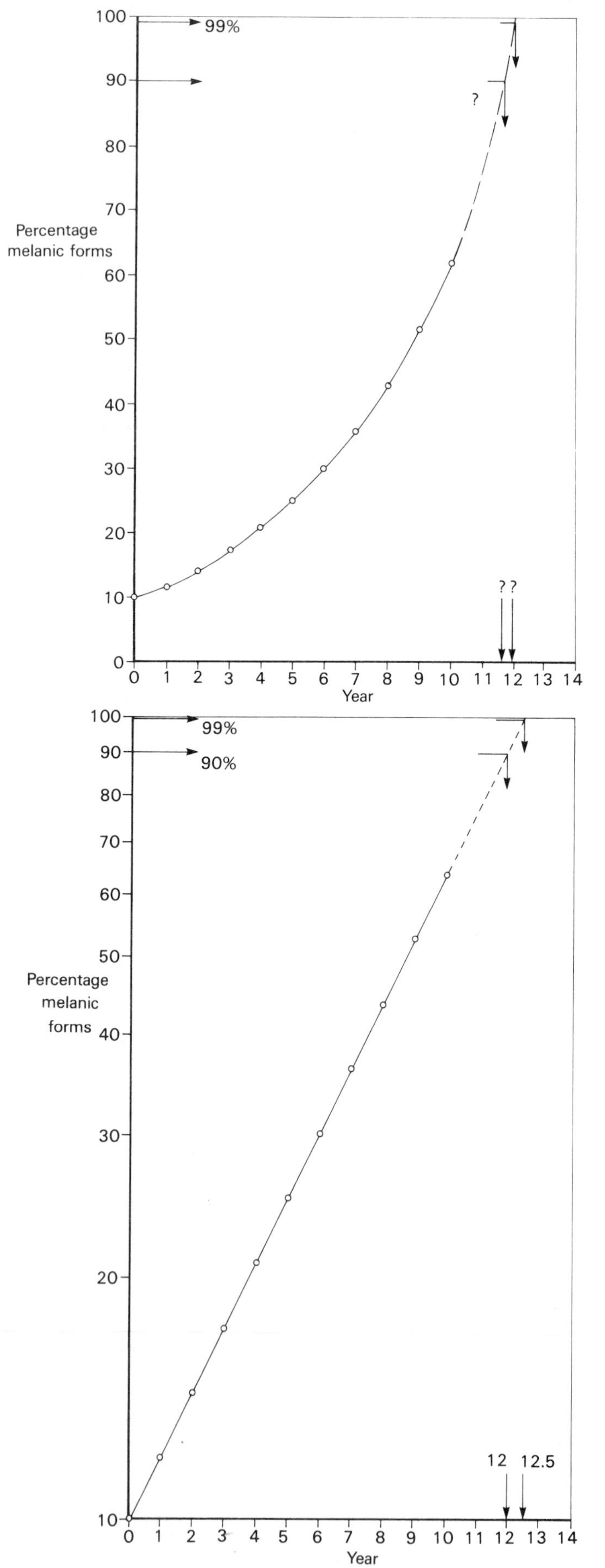

2 (i) 90 per cent: approximately $11\frac{1}{2}$ to 12 years, but it is difficult to determine accurately.

(ii) 99 per cent: approximately 12 years but again it is difficult to determine accurately. Extrapolation is difficult!

3

4 Extrapolation is much more straightforward since the graph line is straight.

(i) 90 per cent: 12 years

(ii) 99 per cent: $12\frac{1}{2}$ years

5 The second method, using the log/linear graph.

3.1.1, p.52, Work Sheet 40

1

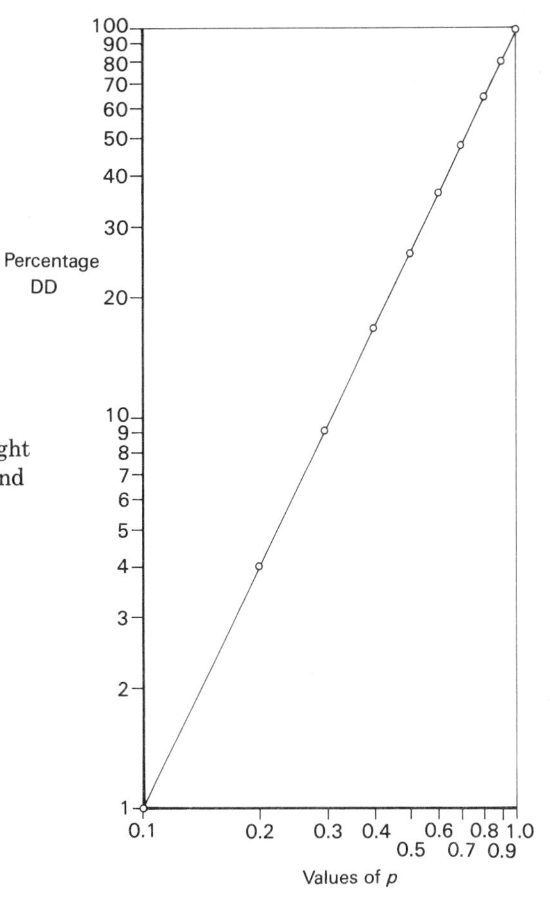

2 Interpolation is more accurate since it is a straight line. Estimating from the graph is thus easier and more accurate.

3.1.2, p.52, Work Sheet 41

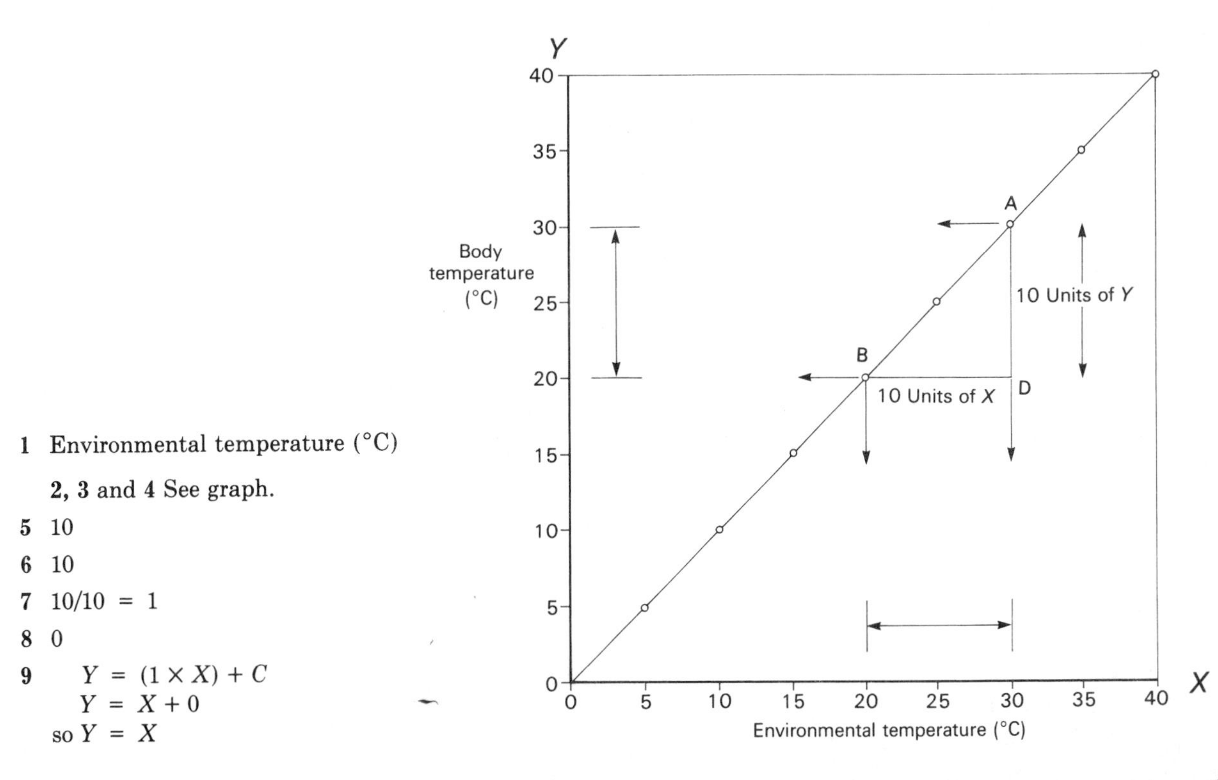

1 Environmental temperature (°C)

2, 3 and 4 See graph.

5 10

6 10

7 10/10 = 1

8 0

9 $Y = (1 \times X) + C$

$Y = X + 0$

so $Y = X$

3.1.2, p.53, Work Sheet 42

1

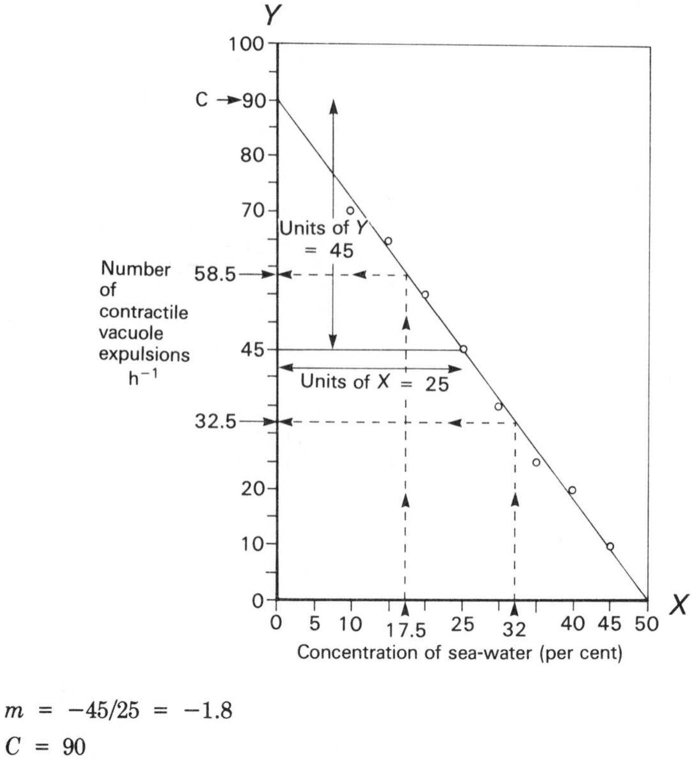

2 $m = -45/25 = -1.8$

3 $C = 90$

4 $Y = (-1.8\,X) + 90$

5 $Y = (-1.8 \times 32) + 90 = 32.4$

6 $Y = 58.5$

7 They should be confirmed.

3.1.2, p.53, Work Sheet 43

1

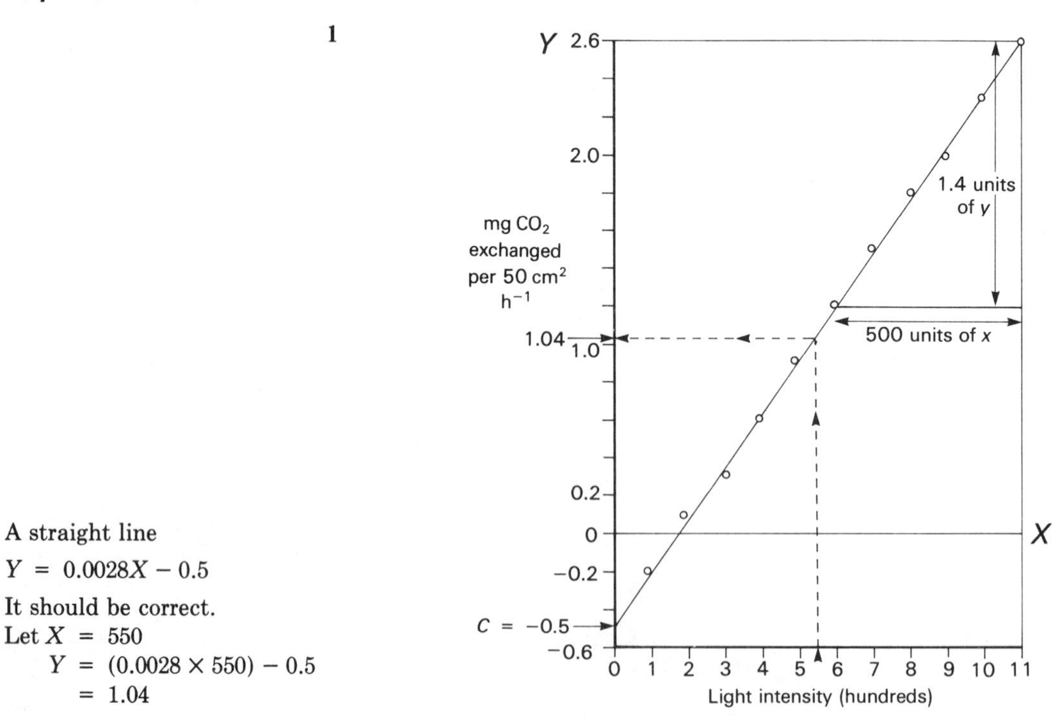

2 A straight line

3 $Y = 0.0028X - 0.5$

4 It should be correct.
Let $X = 550$
$Y = (0.0028 \times 550) - 0.5$
$= 1.04$

3.1.2, p.54, Work Sheet 44

1

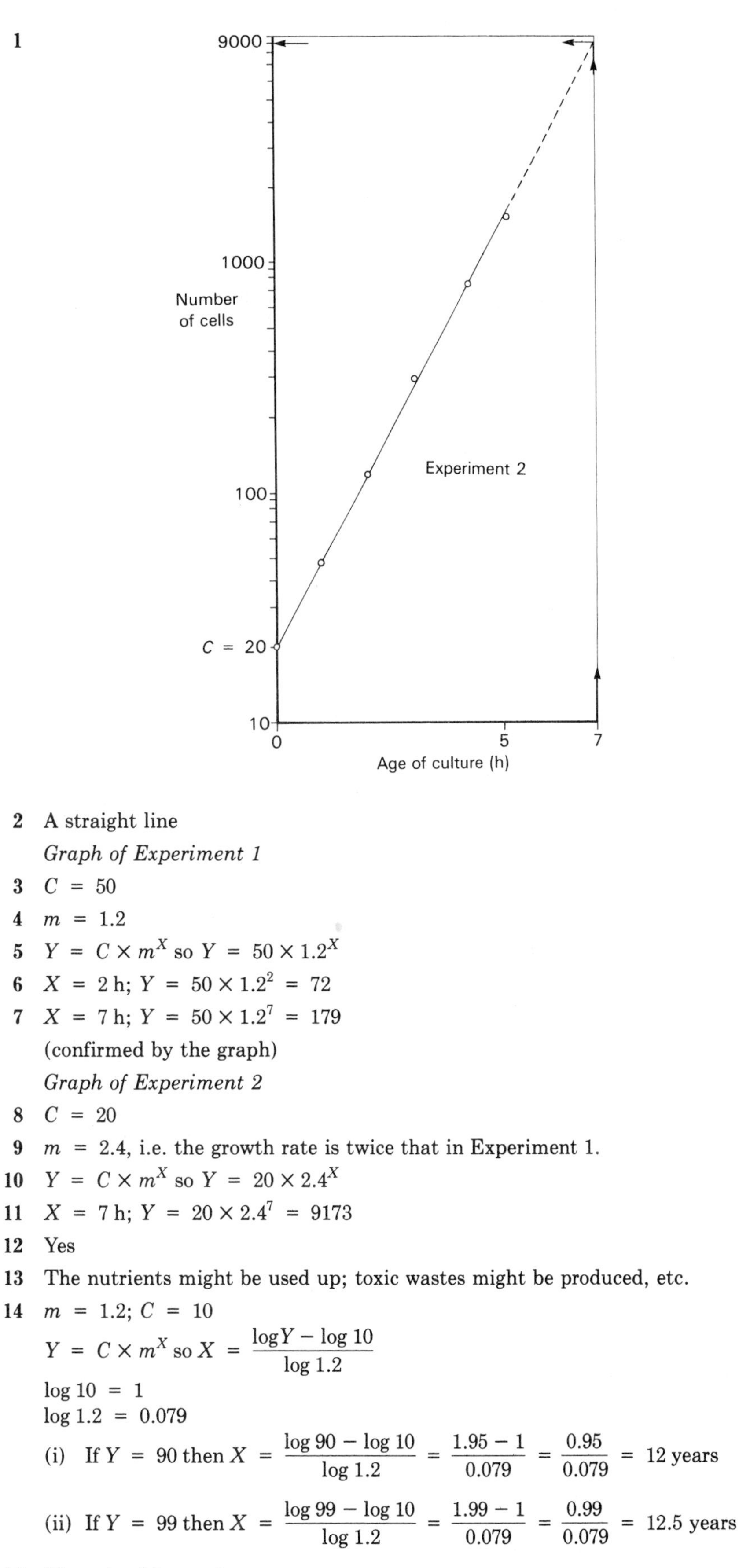

2 A straight line

Graph of Experiment 1

3 $C = 50$

4 $m = 1.2$

5 $Y = C \times m^X$ so $Y = 50 \times 1.2^X$

6 $X = 2$ h; $Y = 50 \times 1.2^2 = 72$

7 $X = 7$ h; $Y = 50 \times 1.2^7 = 179$

(confirmed by the graph)

Graph of Experiment 2

8 $C = 20$

9 $m = 2.4$, i.e. the growth rate is twice that in Experiment 1.

10 $Y = C \times m^X$ so $Y = 20 \times 2.4^X$

11 $X = 7$ h; $Y = 20 \times 2.4^7 = 9173$

12 Yes

13 The nutrients might be used up; toxic wastes might be produced, etc.

14 $m = 1.2$; $C = 10$

$Y = C \times m^X$ so $X = \dfrac{\log Y - \log 10}{\log 1.2}$

$\log 10 = 1$
$\log 1.2 = 0.079$

(i) If $Y = 90$ then $X = \dfrac{\log 90 - \log 10}{\log 1.2} = \dfrac{1.95 - 1}{0.079} = \dfrac{0.95}{0.079} = 12$ years

(ii) If $Y = 99$ then $X = \dfrac{\log 99 - \log 10}{\log 1.2} = \dfrac{1.99 - 1}{0.079} = \dfrac{0.99}{0.079} = 12.5$ years

15 They should agree!

3.1.2, p.55, Work Sheet 45

1

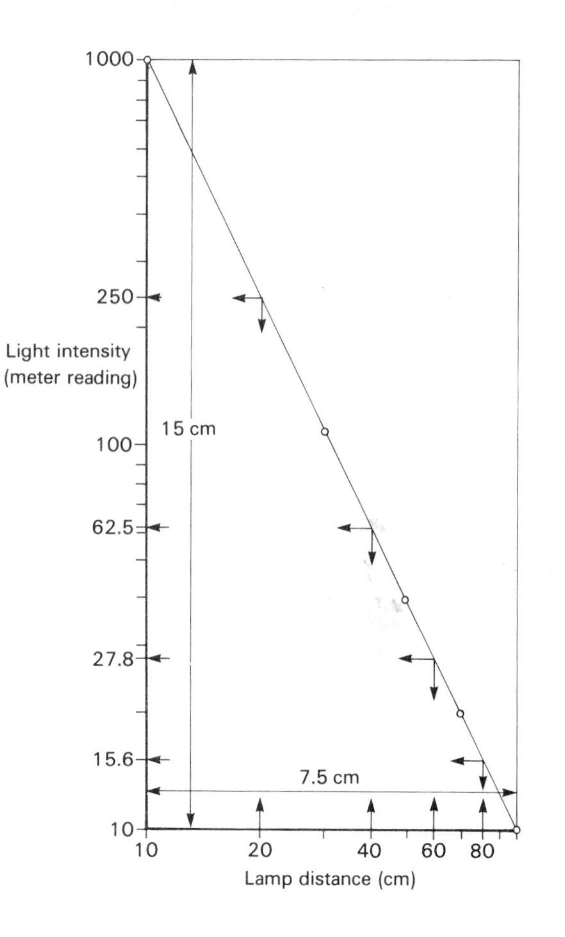

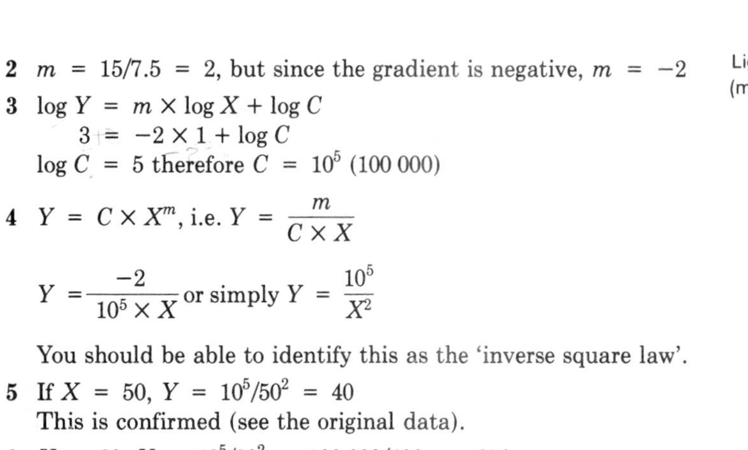

2 $m = 15/7.5 = 2$, but since the gradient is negative, $m = -2$

3 $\log Y = m \times \log X + \log C$
$3 = -2 \times 1 + \log C$
$\log C = 5$ therefore $C = 10^5$ (100 000)

4 $Y = C \times X^m$, i.e. $Y = \dfrac{m}{C \times X}$

$Y = \dfrac{-2}{10^5 \times X}$ or simply $Y = \dfrac{10^5}{X^2}$

You should be able to identify this as the 'inverse square law'.

5 If $X = 50$, $Y = 10^5/50^2 = 40$
This is confirmed (see the original data).

6 $X = 20$, $Y = 10^5/20^2 = 100\,000/400 = 250$
$X = 40$, $Y = 10^5/40^2 = 100\,000/1600 = 62.5$
$X = 60$, $Y = 10^5/60^2 = 100\,000/3600 = 27.8$
$X = 80$, $Y = 10^5/80^2 = 100\,000/6400 = 15.6$

7 All are confirmed by the graph (see details opposite).

3.1.2, p.56

1 Negative 2 $r \simeq -0.7$

3.1.2, p.57, Work Sheet 46

1

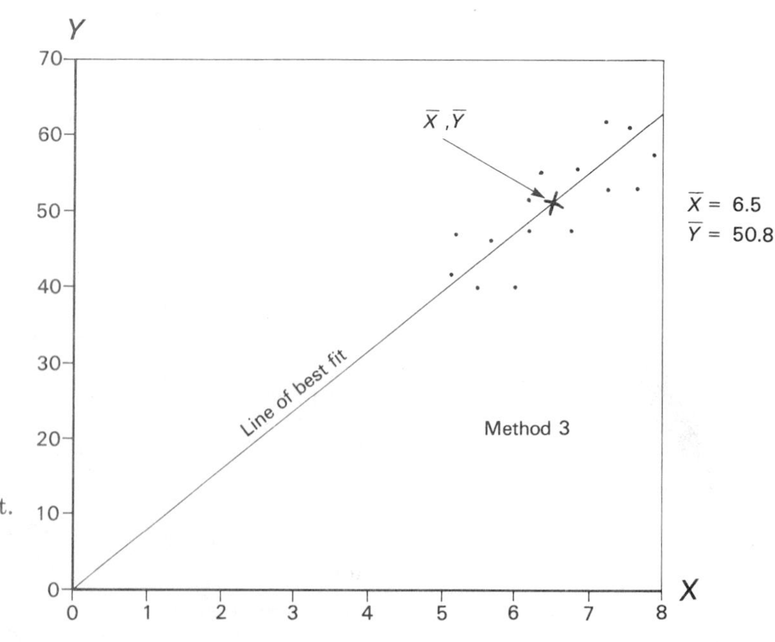

2 It should coincide or at least be close to it.

3 It should
$\overline{X} = 6.5$
$\overline{Y} = 50.8$

3.1.2, p.59, Work Sheet 47

1 and 2

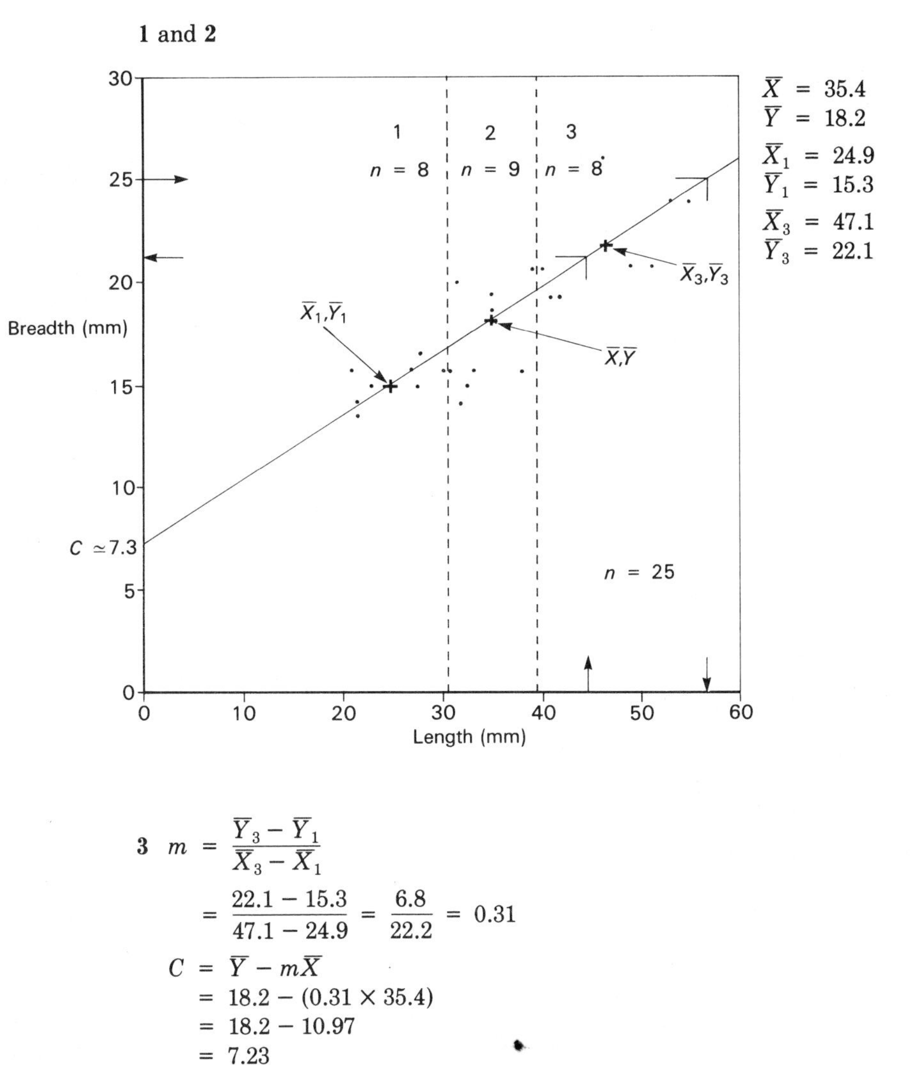

3 $m = \dfrac{\overline{Y}_3 - \overline{Y}_1}{\overline{X}_3 - \overline{X}_1}$

$= \dfrac{22.1 - 15.3}{47.1 - 24.9} = \dfrac{6.8}{22.2} = 0.31$

$C = \overline{Y} - m\overline{X}$
$= 18.2 - (0.31 \times 35.4)$
$= 18.2 - 10.97$
$= 7.23$

On the graph the regression line has been projected to the Y axis to show that this value is confirmed by the graph.

4 The relationship between the length and breadth of this sample of privet leaves is:

$$Y = mX + C, \text{ therefore } Y = 0.31X + 7.23$$

i.e. the breadth (mm) = length (mm) × 0.31 + 7.23 (for this sample)
Checking this equation with the graph:
If the length of a leaf was 45 mm, what would be its breadth?

$$Y = (45 \times 0.31) + 7.23 = 21.2 \text{ (see graph above for confirmation).}$$

If the breadth of a leaf was 25 mm, what would be its length?

$$25 = 0.31X + 7.23$$

$$X = \frac{25 - 7.23}{0.31} = 57.3 \text{ (see graph above for confirmation).}$$

3.1.2, p.60

1 $r = -0.771$

2 It should agree.

3 $r = 0.784$

3.1.4, p.65, Work Sheet 48

1 See graph.
2 Modal class = 39–40 (see graph).
3 Modal value = 39.3 (see graph).

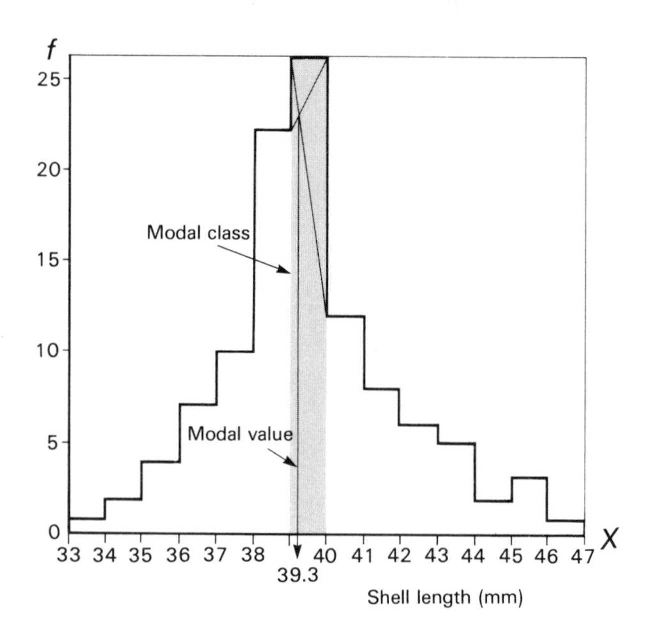

3.1.4, p.72, Work Sheet 49

1 and 2 See graph.
3 2185 [675 + (2 × 500) + (2 × 200) + (2 × 50) + (2 × 5)]
4 1475; 67.5 per cent
5 2080; 95.2 per cent
6 2180; 99.77 per cent
7 Yes, these should be very close.

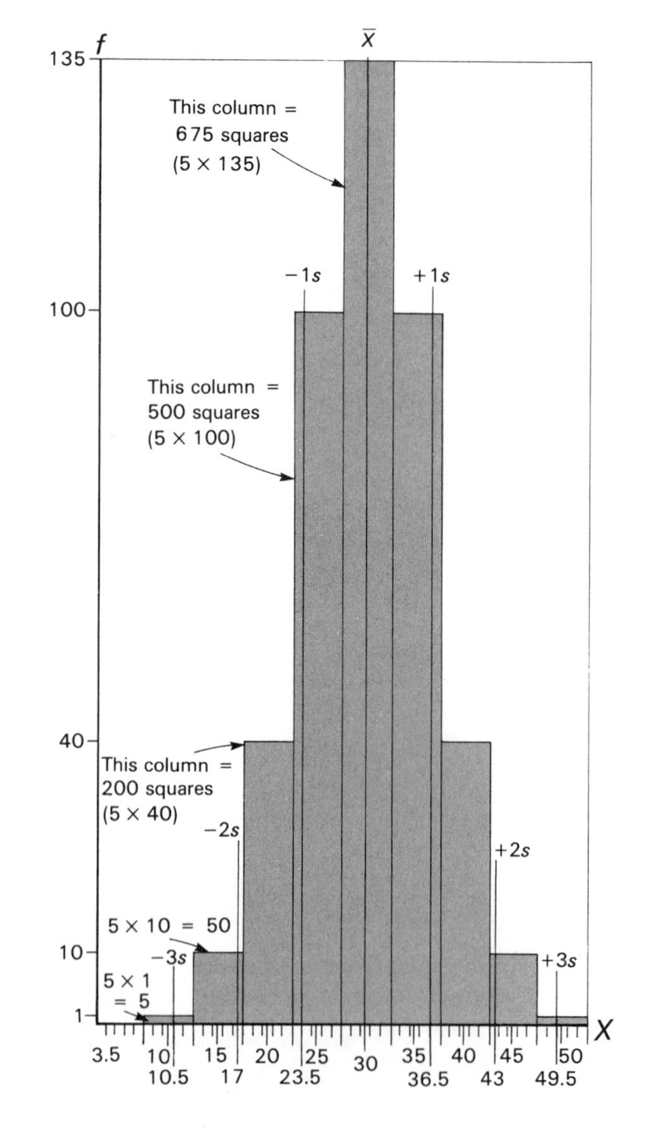

3.1.5, p.73, Work Sheet 50

1

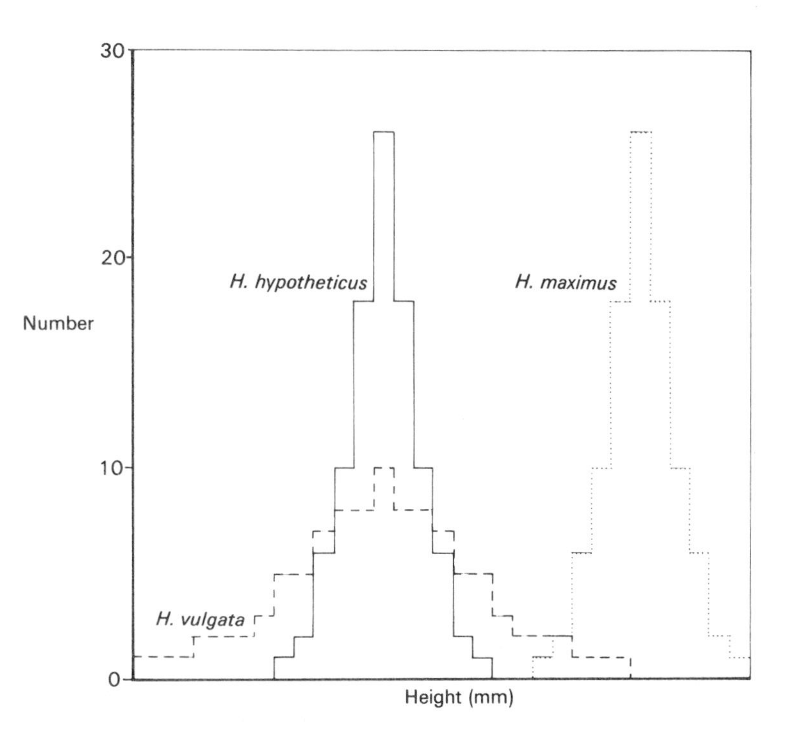

3.1.5, p.73, Work Sheet 51

2

Height	*H. vulgata*		*H. hypotheticus*		*H. maximus*	
(mm)	*f*	Percentage f_{cum}	*f*	Percentage f_{cum}	*f*	Percentage f_{cum}
10–11	1	1				
11–12	1	2				
12–13	1	3				
13–14	2	5				
14–15	2	7				
15–16	2	9				
16–17	3	12				
17–18	5	17	1	1		
18–19	5	22	2	3		
19–20	7	29	6	9		
20–21	8	37	10	19		
21–22	8	45	18	37		
22–23	10	55	26	63		
23–24	8	63	18	81		
24–25	8	71	10	91		
25–26	7	78	6	97		
26–27	5	83	2	99		
27–28	5	88	1	100		
28–29	3	91				
29–30	2	93				
30–31	2	95			1	1
31–32	2	97			2	3
32–33	1	98			6	9
33–34	1	99			10	19
34–35	1	100			18	37
35–36					26	63
36–37					18	81
37–38					10	91
38–39					6	97
39–40					2	99
40–41					1	100

3

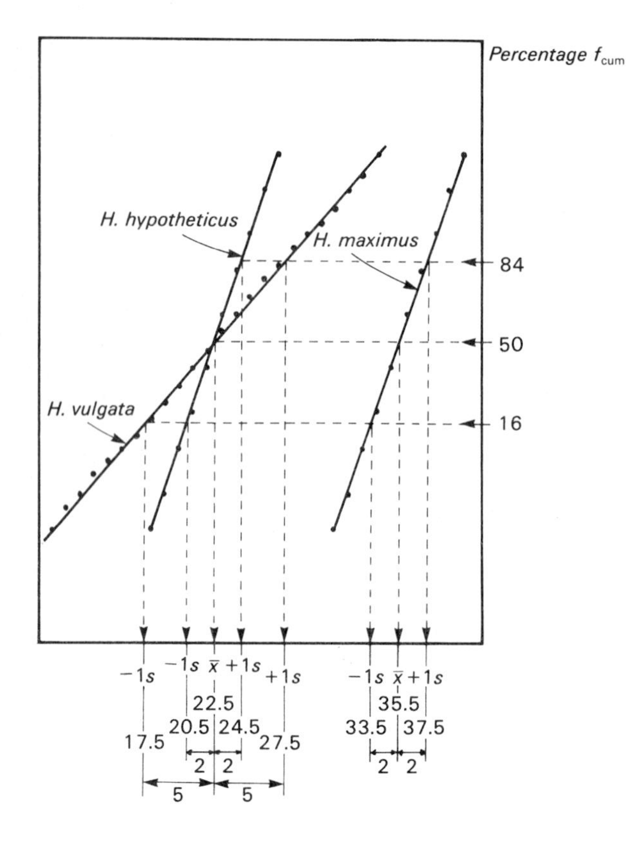

4 The slope of the lines for *H. hypotheticus* and *H. maximus* are the same, and they are both steeper than the line for *H. vulgata*.

If you look at the table and the graph you will see that the spreads of the *H. hypotheticus* and *H. maximus* populations are the same, whereas that for *H. vulgata* is much wider.

The steeper the gradient the smaller the standard deviation.

5 *Means:*

(i) *H. vulgata*, 22.5;
(ii) *H. hypotheticus*, 22.5;
(iii) *H. maximus*, 35.5

From the graph you will see that the means for *H. hypotheticus* and *H. vulgata* are the same, and on the graph below it is the point where the two lines intersect (50 per cent). The mean for *H. maximus* is obviously greater than the other two.

6 *Standard deviations:*

(i) *H. vulgata*, $s = 5$;
(ii) *H. hypotheticus*, $s = 2$;
(iii) *H. maximus*, $s = 2$

From the graph, you will see that the spread of the *H. hypotheticus* and *H. maximus* populations are the same, whereas the spread of the *H. vulgata* population is much greater.

We can therefore use this graphical technique to determine the mean and standard deviation of a population which exhibits a normal, or near normal, distribution.

3.2.1, p.83

1 Percentage of litters having eight piglets or less.

$$Z = \frac{x - \bar{x}}{s} = \frac{8 - 10.2}{2.1} = \frac{-2.2}{2.1} = -1.05$$

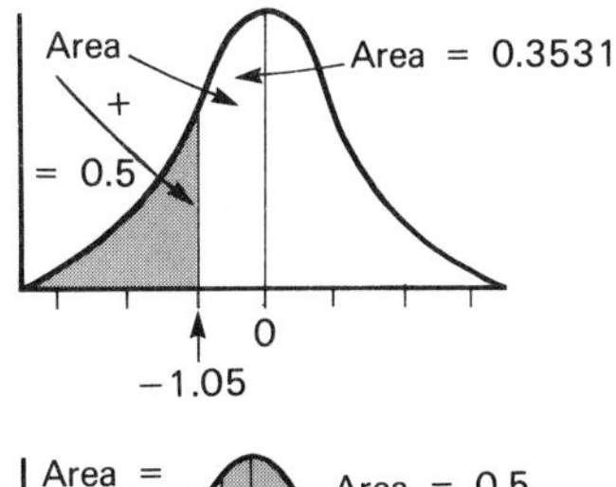

Area $Z = 0$ to $Z = -1.05$ is 0.3531 (from the table); total area to the left of $Z = 0$ is 0.5; therefore the shaded area $= 0.5 - 0.3531 = 0.1469$, so 14.69 per cent of the litters would be expected to have eight piglets or less.

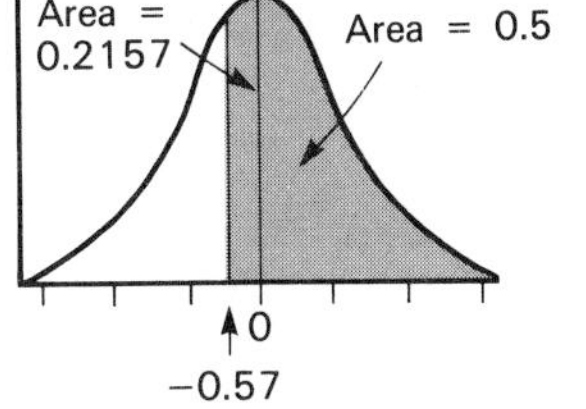

2 Percentage of litters having nine piglets or more.

$$Z = \frac{9 - 10.2}{2.1} = \frac{-1.2}{2.1} = -0.57$$

Area $Z = 0$ to $Z = -0.57$ is 0.2157; total area to the right of $Z = 0$ is 0.5; therefore the shaded area $= 0.5 + 0.2157 = 0.7157$, so 71.57 per cent of the litters would be expected to have nine piglets or more.

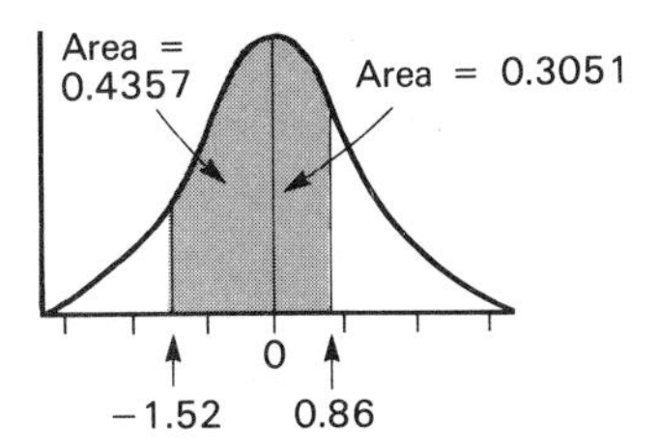

3 Percentage of litters having between seven and twelve piglets.

$$Z \text{ for } 7 = \frac{7 - 10.2}{2.1} = \frac{-3.2}{2.1} = -1.52$$

$$Z \text{ for } 12 = \frac{12 - 10.2}{2.1} = \frac{1.8}{2.1} = 0.86$$

Area $Z = 0$ to $Z = -1.52$ is 0.4357 (from table);
Area $Z = 0$ to $Z = 0.86$ is 0.3051 (from table).
Therefore shaded area $= 0.4357 + 0.3531 = 0.7408$, so 74.08 per cent of the litters would be expected to have between seven and twelve piglets.

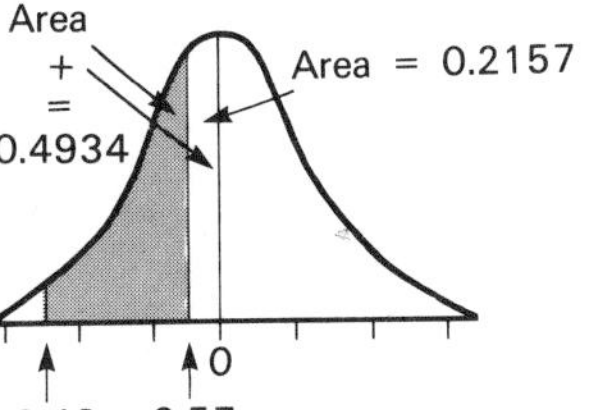
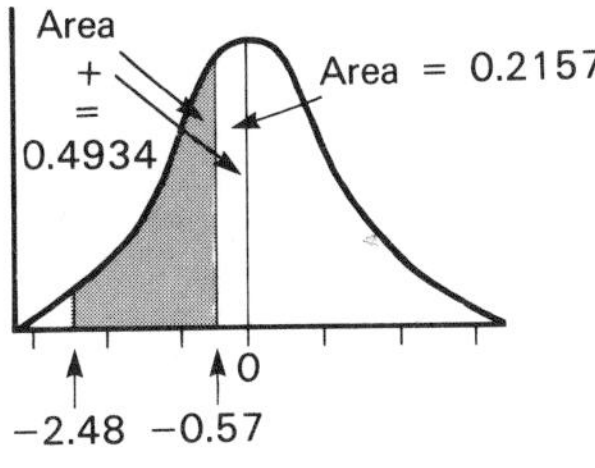

4 Percentage of litters having between five and nine piglets:

$$Z \text{ for } 5 = \frac{5 - 10.2}{2.1} = \frac{-5.2}{2.1} = -2.48$$

$$Z \text{ for } 9 = \frac{9 - 10.2}{2.1} = \frac{-1.2}{2.1} = -0.57$$

Area $Z = 0$ to $Z = -2.48$ is 0.4934 (from table);
Area $Z = 0$ to $Z = -0.57$ is 0.2157 (from table).
Therefore shaded area $= 0.4934 - 0.2157 = 0.2777$ so 27.77 per cent of the litters would be expected to have between five and nine piglets.

3.2.1, p.86, Work Sheet 52

		1	2	
Age of culture (h)	**Mean ($\bar{x}$)**	**Standard deviation (s)**	**95 per cent confidence limits**	
			Upper	**Lower**
0	**50**	**9.7**	**62.2**	**37.8**
1	**60**	**8.7**	**71.1**	**48.9**
2	**72**	**9.4**	**83.9**	**60.1**
3	**86.4**	**14.0**	**104.2**	**68.6**
4	**103.6**	**15.3**	**123.1**	**84.1**
5	**124.4**	**20.6**	**150.5**	**98.3**

2 *Calculation of 95 per cent confidence limits*
Hours

0: $50 \pm 2.78 \times 9.7/2.2 = 50 \pm 2.78 \times 4.4 = 50 \pm 12.2 = 62.2$ and 37.8
1: $60 \pm 2.78 \times 8.7/2.2 = 60 \pm 2.78 \times 4.0 = 60 \pm 11.1 = 71.1$ and 48.9
2: $72 \pm 2.78 \times 9.4/2.2 = 72 \pm 2.78 \times 4.3 = 72 \pm 11.9 = 83.9$ and 60.1
3: $86.4 \pm 2.78 \times 14.0/2.2 = 86.4 \pm 2.78 \times 6.4 = 86.4 \pm 17.8 = 104.2$ and 68.6
4: $103.6 \pm 2.78 \times 15.3/2.2 = 103.6 \pm 2.78 \times 7.0 = 103.6 \pm 19.5 = 123.1$ and 84.1
5: $124.4 \pm 2.78 \times 20.6/2.2 = 124.4 \pm 2.78 \times 9.4 = 124.4 \pm 26.1 = 150.5$ and 98.3

3, 4 and 5

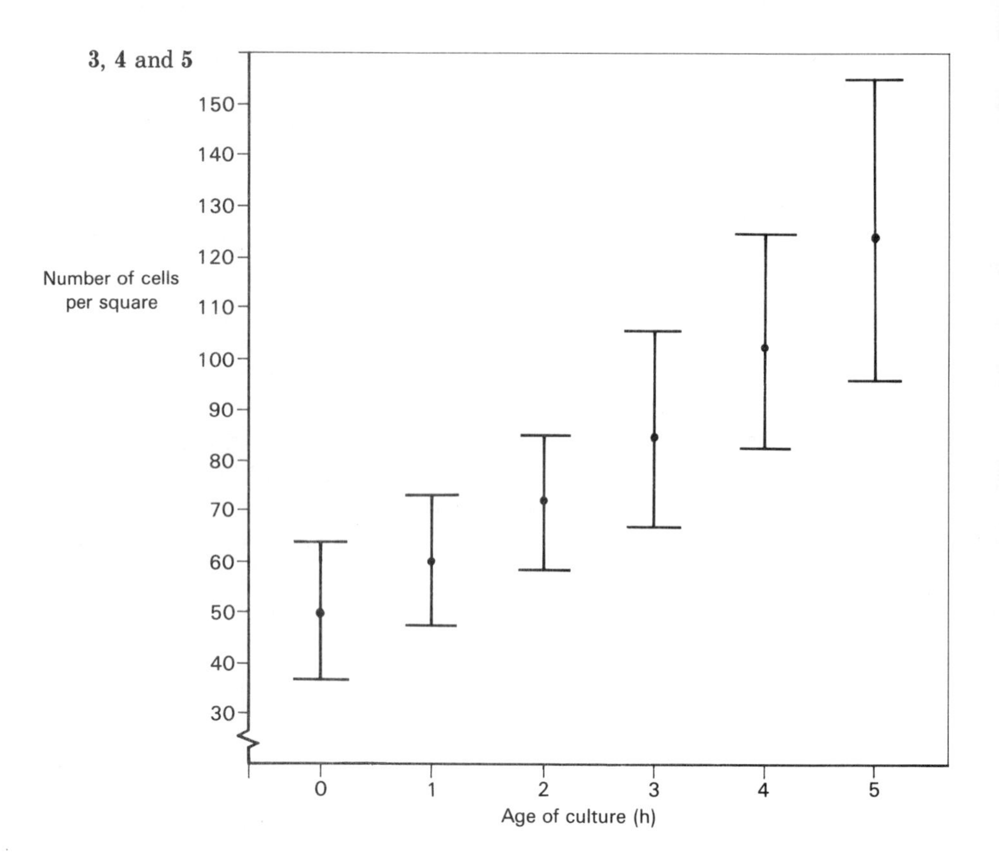

3.2.1, p.86, Work Sheet 53

Colour of light	Mean ($\bar{x}$)	1 Standard deviation (s)	2 95 per cent confidence limits	
			Upper	Lower
Violet	58	7.3	63.2	52.8
Blue	40	5.5	43.8	36.2
Blue-green	62	8.9	68.3	55.7
Green	132	18.7	145.3	118.7
Yellow	96	8.7	102.3	89.7
Orange-red	70	15.3	80.8	59.2

2 Calculation of 95 per cent confidence limits

Violet : $58 \pm 2.26 \times 7.3/3.16 = 58 \pm 2.26 \times 2.3 = 58 \pm 5.2$

Blue : $40 \pm 2.26 \times 5.5/3.16 = 40 \pm 2.26 \times 1.7 = 40 \pm 3.8$

Blue-green : $62 \pm 2.26 \times 8.9/3.16 = 62 \pm 2.26 \times 2.8 = 62 \pm 6.3$

Green : $132 \pm 2.26 \times 18.7/3.16 = 132 \pm 2.26 \times 5.9 = 132 \pm 13.3$

Yellow : $96 \pm 2.26 \times 8.7/3.16 = 96 \pm 2.26 \times 2.8 = 96 \pm 6.3$

Orange-red : $70 \pm 2.26 \times 15.3/3.16 = 70 \pm 2.26 \times 4.8 = 70 \pm 10.8$

3

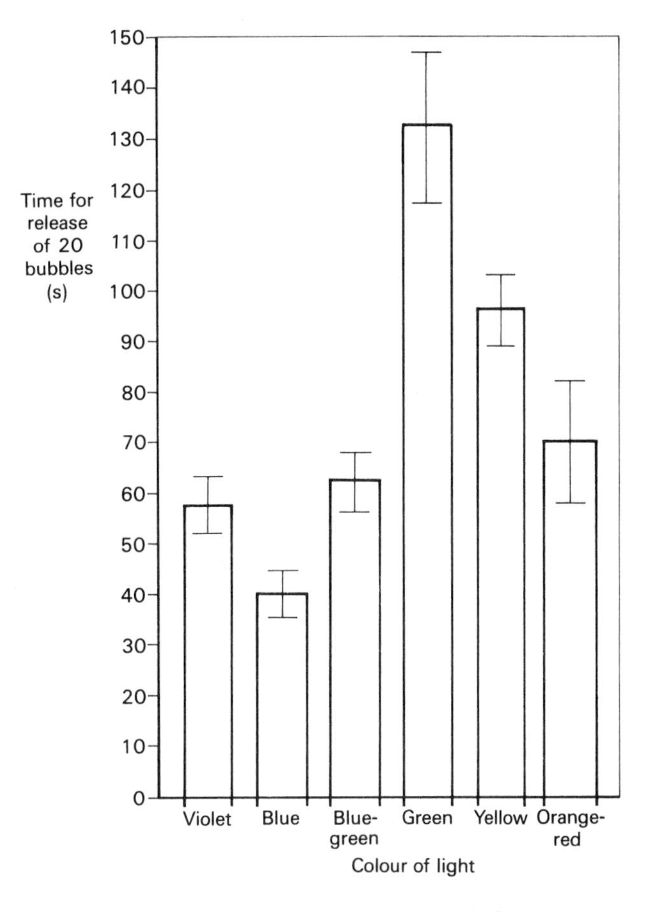

4 Greatest variation, green light; least variation, blue light.

3.2.2, p.90

1 $$\begin{aligned} df &= (n_A - 1) + (n_B - 1) \\ &= (20 - 1) + (28 - 1) \\ &= 19 + 27 \\ &= 46 \end{aligned} \quad \text{or} \quad \begin{aligned} &(n_A + n_B) - 2 \\ &= (20 + 28) - 2 \\ &= 48 - 2 \\ &= 46 \end{aligned}$$

3.2.2, p.91

1 $p < 0.05$

2 $p < 0.001$

3.2.2, p.94

1

	Fertiliser A	Fertiliser B
$\bar{x}$	21	23
s^2	11.56	12.25

$$t = \frac{\bar{x}_A - \bar{x}_B}{\sqrt{\frac{s^2_A}{n_A} + \frac{s^2_B}{n_B}}} = \frac{21 - 23}{\sqrt{0.64 + 0.68}} = \frac{2}{1.15} = 1.74$$

$$df = (n_A - 1) + (n_B - 1) = (18 - 1) + (18 - 1) = 34$$

Using the null hypothesis

(i) The null hypothesis states that there is no difference between the two fertilisers.

(ii) $p > 0.05$

(iii) Accept the null hypothesis — there is no difference between the two fertilisers.

Not using the null hypothesis

(i) $p > 0.05$ (NS)

(ii) The difference between the two fertilisers is not significant.

The claim of the advertiser is therefore not justified.

3.2.2, p.95

1

	B	C	D
A			
B	/		
C	/	/	

2 six t tests, i.e. A × B; A × C; A × D; B × C; B × D; C × D.

3 and 4

		Length	Width
A	$\bar{x}$	8.65	9.10
	s^2	5.90	3.76
B	$\bar{x}$	11.35	7.60
	s^2	10.43	1,93
C	$\bar{x}$	17.1	10.15
	s^2	12.04	5.76

Values of t

Length

	A	B
B	2.11	/
C	6.31	3.83

Width

	A	B
B	2.0	/
C	1.07	2.89

Length

	A	B
B	< 0.05	/
C	< 0.001	< 0.01

Width

	A	B
B	NS	/
C	NS	0.01

5 The cells grow, so the further back from the tip they are the longer they are.

3.2.2, p.97

1

Category	O	E	(O − E)	$(O-E)-0.5$	$[(O-E)-0.5]^2$	$[(O-E)-0.5]^2/E$
Dark half	33	25	8	7.5	56.25	2.25
Light half	17	25	8	7.5	56.25	2.25
Totals	50	50				$\sum = 4.5 = \chi^2$

df = 1
$p < 0.05$

The result is significant. We can be fairly confident that the woodlice prefer the dark side of the choice chamber.

2

Category	O	E	$(O-E)$	$(O-E)-0.5$	$[(O-E)-0.5]^2$	$[(O-E)-0.5]^2/E$
Damp half	31	25	6	5.5	30.25	1.21
Dry half	19	25	6	5.5	30.25	1.21
Totals	50	50				$\sum = 2.42 = \chi^2$

df = 1
$p > 0.10$ NS

There is no significant difference between the numbers in the damp and dry halves of the choice chamber; the woodlice did not show any preference for the damp or the dry half.

3.2.2, p.99

1

Trap	O	E	$(O-E)$	$(O-E)^2$	$(O-E)^2/E$
A	22	23	1	1	0.04
B	26	23	3	9	0.39
C	21	23	2	4	0.17
E	23	23	0	0	0
Totals	92	92			$\sum = 0.6$

$\chi^2 = 0.6$
df = 4 − 1 = 3
$p > 0.5$ (NS)

There is no significant difference between the numbers of field mice caught in the traps.

2 Trap faulty, e.g. door not closing properly; mice able to open the door; some mice entering the trap repeatedly; not so many mice in that part of the forest, etc.

3.2.2, p.100

1

Category	O	E	$(O-E)$	$(O-E)-0.5$	$[(O-E)-0.5]^2$	$[(O-E)-0.5]^2/E$
Inflated pods	882	885.75	3.75	3.25	10.56	0.012
Wrinkled pods	299	295.25	3.75	3.25	10.56	0.036
Totals	1181	1181.0				$\sum = 0.048$

$\chi^2 = 0.048$
df = 1
$p > 0.50$ (NS)

There is no significant difference between these figures and a 3 : 1 ratio; there is no indication of bias on this occasion.

3.2.2, p.100

1

Phenotype	Genotype
Black	AABB (given)
White	AaBb (given
Grey	AABb — strain 1
	AaBB — strain 2
Lethal	double recessives

Cross: white × white

AaBb × AaBb

	AB	Ab	aB	ab
AB	AABB black	AABb grey	AaBB grey	AaBb white
Ab	AABb grey	AAbb X	AaBb white	Aabb X
aB	AaBB grey	AaBb white	aaBB X	aaBb X
ab	AaBb white	Aabb X	aaBb X	aabb X

Ratio : 1 black : 4 grey : 4 white

X = died due to lethal gene combination

3.2.2, p.101

1

Categories	O	E	$(O-E)$	$(O-E)^2$	$(O-E)^2/E$
Agouti	39	36	3	9	0.25
Black	11	12	1	1	0.08
White	14	16	2	4	0.25
Totals	64	64			$\sum = 0.58$

$\chi^2 = 0.58$
df $= 3 - 1 = 2$
$p > 0.05$ (NS)

There is no significant difference between the numbers of offspring and a 9 : 3 : 4 ratio.

3.2.2, p.102

1

Categories	O	E	$(O-E)$	$(O-E)^2$	$(O-E)^2/E$
Yellow, round	**315**	**312.75**	**2.25**	**5.06**	**0.016**
Yellow, wrinkled	**101**	**104.25**	**3.25**	**10.56**	**0.101**
Green, round	**108**	**104.25**	**3.75**	**14.06**	**0.135**
Green, wrinkled	**32**	**34.75**	**2.75**	**7.56**	**0.218**
Totals	556	556.00			$\sum = 0.47$

$\chi^2 = 0.47$
df $= 4 - 1 = 3$
$p > 0.90$ (NS)

There is no significant difference between the observed figures and a 9 : 3 : 3 : 1 ratio – actually they nearly match too closely for comfort!

3.2.2, p.104

1

Student	O	E	$(O-E)$	$(O-E)^2$	$(O-E)^2/E$
1	**68**	**73.4**	**5.4**	**29.16**	**0.397**
2	**76**	**73.4**	**2.6**	**6.76**	**0.092**
3	**64**	**73.4**	**9.4**	**88.36**	**1.204**
4	**48**	**73.4**	**25.4**	**645.16**	**8.790**
5	**78**	**73.4**	**4.6**	**21.16**	**0.288**
6	**82**	**73.4**	**8.6**	**73.96**	**1.008**
7	**71**	**73.4**	**2.4**	**5.76**	**0.078**
8	**65**	**73.4**	**8.4**	**70.56**	**0.961**
9	**110**	**73.4**	**36.6**	**1339.56**	**18.250**
10	**72**	**73.4**	**1.4**	**1.96**	**0.027**
Totals	734	734.0			$\sum = 31.095$

$\chi^2 = 31.095$

For df 9 this gives a p value of less than 0.001, so there is a highly significant difference between the readings.

Looking at the data again we can see that students 4 and 9 diverge widely from the rest. Let us leave them out and try again.

Student	O	E	$(O-E)$	$(O-E)^2$	$(O-E)^2/E$
1	**68**	**72**	**4**	**16**	**0.22**
2	**76**	**72**	**4**	**16**	**0.22**
3	**64**	**72**	**8**	**64**	**0.89**
5	**78**	**72**	**6**	**36**	**0.50**
6	**82**	**72**	**10**	**100**	**1.39**
7	**71**	**72**	**1**	**1**	**0.014**
8	**65**	**72**	**7**	**49**	**0.68**
10	**72**	**72**	**0**	**0**	**0**
Totals	576	576			$\sum = 3.914$

$\chi^2 = 3.914$

For df 7 the value of p is between 0.50 (6.35) and 0.90 (2.83). There is thus no significant difference between these values and they can safely be grouped together.

3.2.2, p.107

1

	Banded	Unbanded	Totals
Broken	68 (63.3)	30 (34.7)	98
Intact	342 (346.7)	195 (190.3)	537
Totals	410	225	635

Cell	O	E	$(O-E)$	$(O-E)-0.5$	$[(O-E)-0.5]^2$	$[(O-E)-0.5]^2/E$
A	68	63.3	4.7	4.2	17.64	0.28
B	342	346.7	4.7	4.2	17.64	0.05
C	30	34.7	4.7	4.2	17.64	0.51
D	195	190.3	4.7	4.2	17.64	0.09
Totals	635	635.0				$\sum = 0.93$

$\chi^2 = 0.93$
df $= (2-1) \times (2-1) = 1$
$p > 0.01$ (NS)

There is no relationship between banding and predation.

2 Using the alternative formula of χ^2 for 2×2 contingency tables:

$$\chi^2 = \frac{n([ad - bc] - \frac{1}{2}n)^2}{(a+b)(c+d)(a+c)(b+d)} = \frac{635([13\,260 - 10\,260] - 317.5)^2}{92\,250 \times 52\,626}$$

$$= \frac{635\,(3000 - 317.5)^2}{92\,250 \times 52\,626} = \frac{635 \times 7\,195\,806}{92\,250 \times 52\,626} = 0.94$$

3.2.2, p.108

1

Dandelions	Daisies		
	Absent	Present	Totals
Present	10 (22.5)	40 (27.5)	50
Absent	35 (22.5)	15 (27.5)	50
Totals	45	55	100

Cell	O	E	$(O-E)$	$(O-E)-0.5$	$[(O-E)-0.5]^2$	$[(O-E)-0.5]^2/E$
A	10	22.5	12.5	12	144	6.4
B	35	22.5	12.5	12	144	6.4
C	40	27.5	12.5	12	144	5.2
D	15	27.5	12.5	12	144	5.2
Totals	100	100.0				$\sum = 23.2$

$\sum^2 = 23.2$
df $= 1$
$p < 0.001$

The result is very highly significant – we can be almost certain that there is an association between the daisies and the dandelions on this lawn.

2 The association is positive – wherever daisies occur so also do dandelions. Hypotheses could be proposed to explain why this should be so and suitable experiments devised to test them.

3.2.2, p.111

1

O	E	$(O - E)$	$(O - E)^2$	$(O - E)^2/E$
97	83.1	13.9	193.2	2.33
16	20.4	4.4	19.36	0.95
17	14.5	2.5	6.25	0.43
12	16.4	4.4	19.36	1.18
10	17.6	7.6	57.76	3.28
144	147.6	3.6	12.96	0.09
38	36.3	1.7	2.89	0.08
22	25.7	3.7	13.69	0.53
32	29.2	2.8	7.84	0.27
34	31.2	2.8	7.84	0.25
20	25.1	5.1	26.01	1.04
7	6.2	0.8	0.64	0.10
5	4.4	0.6	0.36	0.08
6	5.0	1.0	1.00	0.20
8	5.3	2.7	7.29	1.38
32	37.2	5.2	27.04	0.73
11	9.1	1.9	3.61	0.40
7	6.5	0.5	0.25	0.04
8	7.4	0.6	0.36	0.05
10	7.9	2.1	4.41	0.56

$\sum = 13.97$

$\chi^2 = 13.97$

2 df $= (c - '1)(r - 1)$
$= 3 \times 4 = 12$
$p > 0.1$ (NS)

There is no relationship between eye colour and colour preference.

3.2.2, p.112

1 $r = -0.771$
df $= 30 - 2 = 28$
$p < 0.001$

The correlation is very highly significant and negative.

2 $r = 0.784$
df $= 13$
$p < 0.001$

The correlation is very highly significant and positive.

3 $r = 0.866$
df $= 23$
$p < 0.001$

The correlation is very highly significant – we can be almost certain that there is a correlation between the length and the breadth of privet leaves.

(*WS* = Work Sheets)